THOMAS CRANE PUBLIC LIBRARY
QUINCY MA

CITY APPROPRIATION

FACTS AT YOUR FINGERTIPS
ECOLOGY

Published by Brown Bear Books Limited

4877 N. Circulo Bujia
Tucson, AZ 85718
USA

and

First Floor
9-17 St. Albans Place
London N1 0NX
UK
www.brownreference.com

© 2010 The Brown Reference Group Ltd

Library of Congress Cataloging-in-Publication Data

Ecology / edited by Sarah Eason.
    p. cm. – (Facts at your fingertips)
  Includes index.
  ISBN 978-1-936333-04-2 (library binding)
  1. Ecology–Juvenile literature. I. Eason, Sarah. II. Title. III. Series.

QH541.14.E242 2010
577-dc22

2010015184

All rights reserved. This book is protected by copyright. No part of it may be reproduced, stored in a retrieval system, or transmitted in any form or by any means, without the prior permission in writing of the Publisher, nor be otherwise circulated in any form of binding or cover other than that in which it is published and without a similar condition including this condition being imposed on the subsequent publisher.

ISBN-13 978-1-936333-04-2

Editorial Director: Lindsey Lowe
Editor: Sarah Eason
Proofreader: Jolyon Goddard
Designer: Paul Myerscough
Design Manager: David Poole
Children's Publisher: Anne O'Daly
Production Director: Alastair Gourlay

Printed in the United States of America

Picture Credits

Abbreviations: b=bottom; c=center; t=top; l=left; r=right.

Front Cover: Shutterstock: Reef
Back Cover: Shutterstock: Rich Carey

Dreamstime: Dsabo 48, Simon Whitehouse 57; **Fotolia:** MiklG 45, Outdoorsman 5; **istockphoto:** Michael Braun 22, Carl Chapman 8, Eric Delmar 49, Steve Estvanik 42, Michael Hieber 20, Luoman 47; **Shutterstock:** Galyna Andrushko 35, Basel101658 12, Magdalena Bujak 11, Sascha Burkard 58, Rich Carey 41, Lucian Coman 27, Condor36 32, Brittany Courville 40, Lawrence Cruciana 60, David Davis 46, Pichugin Dmitry 15t, Epic Stock 38, Mike Flippo 61, FloridaStock 3, Arto Hakola 6, David Hancock 1, Daniel Hebert 17, Andy Heyward 36, Isantilli 21, Eric Isselée 10, Gail Johnson 15b, Peter Kirillov 52, Arnold John Labrentz 39, Lockenes 33, MDD 37, Natalia Sinjushina & Evgeniy Meyke 13, Tan Wei Ming 56, Morgan Lane Photography 54, Orionmystery 50, Dr. Morley Read 34, Graeme Shannon 24, Smit 19, Andrey Tarantin 30, Voyagerix 4, Bershadsky Yuri 16, A.S. Zain 55.

Artwork © The Brown Reference Group Ltd

The Brown Reference Group Ltd has made every effort to trace copyright holders of the pictures used in this book. Anyone having claims to ownership not identified above is invited to contact The Brown Reference Group Ltd.

CONTENTS

What is Ecology?	4
The Diversity of Life	8
Cycles of Nature	16
Climate and Earth	22
Ecosystems	30
Watery World	38
Human Impact	46
Conservation	56
Glossary	62
Further Resources	63
Index	64

WHAT IS ECOLOGY?

Ecology is about the pattern of nature—it is the study of the many interactions among living organisms and their environment.

Every living thing depends on other things for its survival. House sparrows living in a park have to find seeds to eat. The seeds come from plants that must find places to grow. Sparrows feed on insects too, and the insects need plants to eat. Sparrows also need air to breathe, water to drink, and places to lay eggs. So each sparrow is at the center of a web of relationships involving other living things and its surroundings, or environment. Ecology is the science that studies these types of relationships. Instead of concentrating only on a sparrow, ecologists study how it interacts with other organisms and its environment.

Many caterpillars eat leaves, and many birds then eat the caterpillars. This link between leaf, caterpillar, and bird forms a food chain.

Food chains and webs

One of the most basic ideas in ecology is the **food chain**. Green plants make food from water, sunlight, and a gas called carbon dioxide (in the air) by a process called **photosynthesis**. Animals cannot do this, so they must eat other living organisms to get their food.

Caterpillars eat leaves and turn them into caterpillar flesh. In a simple food chain other animals, such as small birds, eat the caterpillars, and the birds themselves may be eaten by cats. However, the caterpillars are also eaten by other insects. Thus most food chains are not so simple. Several linked chains make up a **food web**.

FOOD WEBS

This diagram shows a food web based on the plants and trees in a small forest. A food web is much more complicated than a simple food chain, in which each plant or animal provides food for just one other animal. The plants provide food for many insects, worms, and other small animals, which in turn provide food for larger animals, such as cats, moles, and many different types of birds.

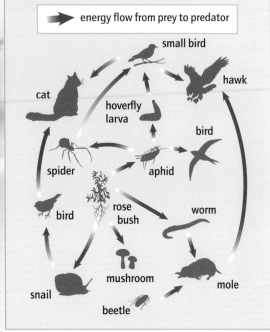

ECOLOGY

TRY THIS

Compile a Food Web

Make a list of all the animals that live in your backyard or in a local park. Remember to include insects and other minibeasts, and all the animals with fur, feathers, or scales. Check out the animals in a book to see what they eat. Then try to arrange them in a food web like the one shown on page 4. Be warned—it might get complicated!

Habitats and niches

All animals, plants, and other living organisms struggle to survive. Many organisms die, and only the best adapted live long enough to reproduce. This process is called **natural selection**, the basis of evolution. What does "fittest" mean? That depends on the environment. A goldfish can survive very well in a pond, but it would die in a desert. A lizard can live in a desert, but it would freeze in the Arctic. Even if an animal or plant survives, it may not do as well as its neighbors. So over time it is crowded out. Every living organism has characteristics, or adaptations, that make them suited to their environment. The place in which an animal or plant lives is called its **habitat**.

A habitat could be a rocky seashore or a tropical forest. Such places offer all kinds of ways in which animals and plants can live. Every **species** of living organism has its own special way of surviving in its habitat, and that is called its **niche**. For example, some birds specialize in eating large fruits, while others feed on small insects. The two types of birds occupy different niches.

Population to community

A niche can be occupied by just one species. A small lake may have only enough food for one big predatory fish, such as a pike, but a large lake contains enough food for several pike, and they form a population. All the animals in a population are of the same species and occupy the same ecological niche. They share their lives with other populations of different animals and plants, and together these interacting populations form a **community**. So the fish, insects, birds, and plants that live in or around a lake make up a community.

With their thick, white fur, these polar bears are perfectly adapted to life in the frozen Arctic.

WHAT IS ECOLOGY?

Ecosystems

An ecological community can contain many organisms including plants, animals, fungi, and microorganisms, such as bacteria. These organisms share an environment that has nonliving elements, such as the **climate** (the typical weather experienced over a year), the soil, and the location, which might be an exposed, rocky headland or a sheltered, sandy beach. A river might be slow and muddy or swift and sparkling. The possibilities are endless.

Communities of living organisms interact with their nonliving environment in all kinds of ways. These interactions often control the types of living things present in the community. For example, most riverside plants cannot grow in an estuary (where the tide mixes with freshwater) because the water is too salty. Only salt-tolerant plants can grow in estuary shallows. Along with the animals that live among them these plants form a salt-marsh community. The cycle of complex interactions between the community and its environment is called an **ecosystem**.

TRY THIS

Make Your Own Ecosystem

Create your own mini-ecosystem in a big glass jar by collecting some water from a pond in summer. Tell an adult before you go, and take care not to fall in! Fill your jar about three-quarters full. Then add some mud taken from the bottom of the pond. A layer one-to-two fingers thick should be enough. Add some water plants to produce oxygen. You can get them from the pond or at pet stores. Put your jar in a cool window, and wait for a while. The water will clear, and you may be surprised at what you see.

An alligator living in a swamp would be affected if other animals or plants were removed from its habitat.

ECOLOGY

Biomes

Any community in its environment can be called an ecosystem if it looks after itself. A small pond is an ecosystem because all the animals and plants in it get almost everything they need without leaving the water. The plants make the oxygen and food that the animals need, and in turn the animals make the nutrients and carbon dioxide that the plants need. A forest could also be seen as an ecosystem. In some parts of the world forests cover vast areas. So do grasslands, oceans, and deserts. A large geographical region that has its own distinctive climate, plants, and animals is called a **biome** and is generally made of several types of ecosystems. All the deserts on Earth form one biome; all the tropical forests form another biome.

> **SCIENCE WORDS**
> - **climate** The regular weather pattern that occurs in a certain region.
> - **food chain** The passage of energy between organisms; a plant links to a herbivore, which in turn links to a carnivore. Energy is lost with each step.
> - **food web** A complex series of interlinked food chains.
> - **microhabitat** A small part of a habitat that sustains a community; for example, a pool in the leaves of bromeliad plants forms a microhabitat.
> - **niche** The ecological role of an organism in a community.

MICROHABITATS AND COMMUNITIES

Most habitats are made up of much smaller **microhabitats**, such as the forest floor, a hole in a tree, or the different layers of a tree. A forest usually has many different types of trees, and each tree can support its own community of animals. Different communities also live at different levels of the forest. One community lives on the forest floor, another in the understory, and yet another among the branches of the forest (top **canopy**, or branchy layer). Tropical forests have two or three canopy layers. More layers produce more communities. More communities create greater **biodiversity** (range of life-forms).

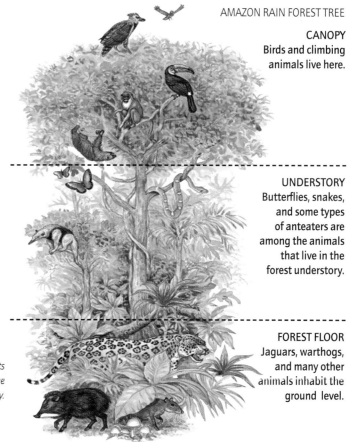

AMAZON RAIN FOREST TREE

CANOPY Birds and climbing animals live here.

UNDERSTORY Butterflies, snakes, and some types of anteaters are among the animals that live in the forest understory.

FOREST FLOOR Jaguars, warthogs, and many other animals inhabit the ground level.

Different types of animals live in different parts of the forest—from the forest floor to the branchy treetops, or top canopy.

7

THE DIVERSITY OF LIFE

Biologists have now identified more than 1.7 million different species (types) of living things, from microscopic bacteria to gigantic blue whales.

Most biologists, however, suspect that the true number of living organisms is many more than 1.7 million. Indeed, 10,000 new species are discovered and named each year—mostly insects and other tiny creatures.

Arranging the diversity of life into some kind of order is a task that has occupied biologists ever since Swedish botanist Carolus Linnaeus (1707-1778) introduced the basic unit for classifying life that remains in use today—the species. A species is a particular kind of living thing, such as a leopard, a date palm, or a woodchuck.

Although every individual animal or plant of a species is different, the members of a species are more like one another than they are any other living thing. Also, they are generally only able to breed with others of the same species. To better understand the relationships between species, biologists divide species into groups, a process called **classification**.

All living things belong to one of the five kingdoms of life: plants, animals, fungi, **protists**, or bacteria. The first four kingdoms are grouped into a domain called the Eukaryota. All eukaryotes consist of cells that contain a membrane-enclosed nucleus—the nucleus contains genetic information that drives the way a cell develops. Eukaryote cells also contain miniorgans called **organelles**. Each of the organelles performs an important task for the cell.

Bacteria are classed in a separate domain, Prokaryota. A prokaryote cell contains neither a membrane-enclosed nucleus nor organelles.

Coral reefs are home to an amazing diversity of life. Coral reefs occupy less than 1 percent of the world's ocean surface, but they are home to more than 25 percent of all marine animals.

ECOLOGY

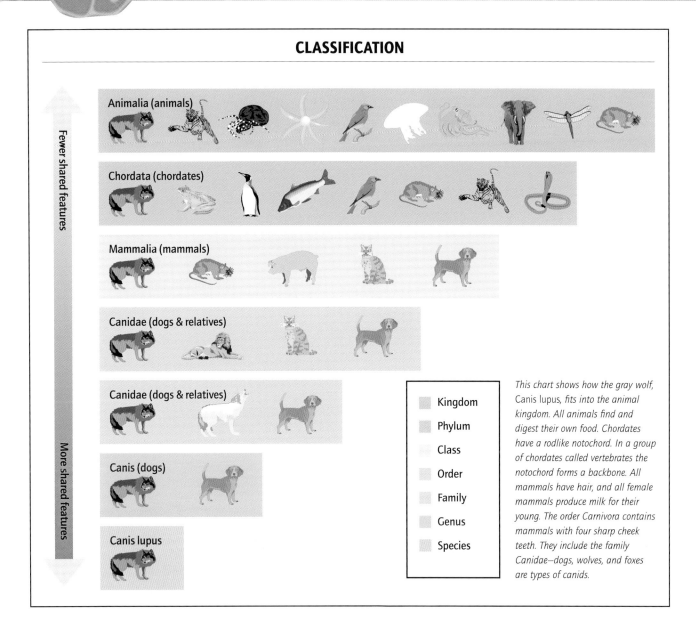

This chart shows how the gray wolf, Canis lupus, fits into the animal kingdom. All animals find and digest their own food. Chordates have a rodlike notochord. In a group of chordates called vertebrates the notochord forms a backbone. All mammals have hair, and all female mammals produce milk for their young. The order Carnivora contains mammals with four sharp cheek teeth. They include the family Canidae—dogs, wolves, and foxes are types of canids.

Higher levels

Related species are grouped into a genus (plural genera). Closely related genera are further grouped into families. For example, all cats belong to one family, the Felidae. Families are placed together into orders, such as the Carnivora, which includes the cat family, as well as all other mammals, such as dogs, that have four sharp cheek teeth (the word **carnivore** refers to all meat-eating animals, but not all are within the order Carnivora.) Orders are then grouped into classes, such as the Mammalia, or mammals.

Classes of plants are grouped into divisions, while classes of animals are grouped into phyla (singular phylum). Each of the phyla includes organisms that share certain features. Phylum Arthropoda, for example, includes at least one million species of invertebrates (animals that do not have a backbone), including spiders, insects, and crabs. Although they may look very different, all arthropods have a hard outer covering, which is called an exoskeleton. They also all have jointed appendages that form legs, mouthparts, and antennae.

THE DIVERSITY OF LIFE

Exceptions to the rule

Not all organisms fit neatly into the order–class–phylum system. For many animal groups, such as insects and snakes, intermediate levels such as superclasses and suborders are created. New discoveries force biologists to rethink classifications, adding new levels or moving a species to a different group. Even the established system that follows the hierarchy of small groups within larger groups may be changed.

Cladistics

In 1950, German insect specialist Willi Hennig (1913–1976) introduced the idea of **cladistics**. The word *cladistics* comes from a Greek word meaning "branch," and cladistics involves the construction of branching networks that trace the ancestry of a species.

The cladistics system compares large numbers of characteristics among species to build up a family tree for a range of species. This tree is called a cladogram. Cladograms show points in the evolution of the group when certain key characteristics emerged for the first time. Within a cladogram there are sets of branches

NAMING AND CLASSIFYING SPECIES

Early naturalists could see that some life-forms were similar and others were not. But until Linnaeus devised his system for naming and classifying life-forms, there was no logical system to arrange them. Linnaeus introduced the use of two-part names for every species and put each species in a series of ever larger groups called a hierarchy.

Every species has a two-part name in Latin. The first part is a genus name shared with closely related species. The second part, the specific name, refers to just one species. The scientific name of the leopard is *Panthera pardus*. *Panthera* is the name of the genus that includes most of the big cats, while *pardus* specifies the leopard alone.

The black panther is a color variant of the leopard; they are the same species. Black jaguars are also called black panthers.

ECOLOGY

called clades. Each clade begins with one common ancestor, and then branches out into all of its descendants, which share the same key feature or features. A branch showing leopards, domestic cats, and all their relatives descending from a common ancestor would form a clade. A group of branches showing that carnivores such as cats, dogs, and bears descend from a common ancestor is another. Unlike biologists making traditional family trees, biologists using cladistics do not look for "shared primitive characters," which are common traits that date back long before the group being studied developed. One such character is a mammal's backbone, a feature that also occurs in animals such as fish and reptiles. Instead, cladistics focuses on when a new feature that defines the group first appeared. This feature is called a "shared derived character." In mammals hair is a shared derived character. So, a biologist constructing a cladogram figures out the point at which hair first appeared and starts the mammal clade from there.

What is a species?

A species is generally defined as a group of organisms that can breed with each other. The word *species* comes from a Latin term meaning "appearance," and biologists originally identified species by the way they looked. A species was simply a group of animals that looked the same. Looking at an organism's shape and form, or morphology, is still the easiest way to identify many species. You can distinguish a red fox from other kinds of dogs by its bushy, red tail, for example.

Appearances can be deceptive, however. There are several populations of plains zebras living in different parts of Africa. They animals differ in appearance but do not form separate species because the different populations can interbreed. Populations like these are called **subspecies**.

Other animals look similar because they have evolved to suit similar habitats. This is called convergent evolution.

The red fox is the wild relative of the domestic dog. Both are members of the family Canidae.

HOW DO YOU DEFINE A SPECIES?

In 1942, German biologist Ernst Mayr (1904-2005) suggested that species are populations of organisms whose individuals can breed with each other. So, a lion can breed with other lions but cannot breed with a hyena. This explains why species remain separate and clearly defined. However, this concept does not include species that do not reproduce sexually, that is, by exchanging genes. Bacteria, for example, only sometimes exchange genes. Also, different species do sometimes breed to create hybrids, such as wheat, a hybrid of two species of plants.

THE DIVERSITY OF LIFE

FUNGI

Mushrooms, toadstools, yeasts, mildews, and the molds that grow on bread, all belong to a group of organisms called fungi. Fungi form their own separate kingdom, containing at least 70,000 different species. They are not plants, and they have no chlorophyll (the green pigment that absorbs energy from sunlight). Genetic tests suggest they are more closely related to animals than to plants, but fungi are not animals either.

Because they cannot make their own food as plants do, fungi eat plants and animals. Parasitic fungi feed on living organisms. Saprophytic fungi feed on plant and animal remains. Both types eat by releasing chemicals called enzymes that digest what the fungi are feeding on. The fungi absorb nutrients and minerals from their food. Chemicals released by some mold fungi give blue cheese its distinctive flavor.

Fungi are made of numerous cottonlike threads called hyphae, which absorb nutrients. Hyphae spread out in a tangled mass through the soil in ground-based fungi or within the tissues of the plant or animal host. The mass of hyphae is called a mycelium. Sometimes hyphae can clump together to form fruiting bodies such as toadstools, or they form pinheads, such as the mold on rotting fruit.

The bright red cap of the fly agaric warns animals that this mushroom is poisonous.

The variety of life

Life on Earth is astonishingly diverse, with up to several millions of different species in the oceans, in the soil, and on land. According to the World Wildlife Fund for Nature's Living Planet Index, there are about 300,000 species of plants, 4,000 species of mammals, 6,300 species of reptiles, 4,200 species of amphibians, 19,100 species of fish, more than 1 million species of insects, and about 400,000 other species of invertebrates, including roughly 35,000 species of spiders. These are merely the numbers of species known to science. Biologists think the actual number of species may be more than 5 million, while some estimates are as high as 100 million species!

When life on Earth began, there were just bacteria-like organisms. All the countless species that have lived in the past and are alive now have gradually evolved from these life-forms.

How new species form

New species emerge when one species splits into two or more populations. The groups may be separated by physical barriers, such as oceans, mountain ranges, or rivers, or by their behavior—perhaps one group is active at dawn and the other

ECOLOGY

at dusk. If the groups are separated for long enough, natural variations between the populations will then eventually lead to their being unable to interbreed. A new species has thus formed. This gradual process is called speciation.

The biodiversity crisis

In recent years the number of species, or biodiversity, has suddenly taken a dramatic downturn due to the activities of people. Today there are more than 5,000 species in the "at risk" category, and many more are under threat. Even optimistic estimates suggest a quarter of all mammal and amphibian species, 11 percent of birds, 20 percent of reptiles, and 34 percent of fish will be in danger of extinction by the year 2020. There may also be a loss of up to 47 percent of all known plant types—a total of at least 144,000 species.

> ### TARGETING HOT SPOTS
>
> In recent years some biologists have focused on the idea of biodiversity hot spot. They are small areas of the world where a huge number of species are concentrated, such as Madagascar, Costa Rica, and the Philippines.
>
> Research by environmental scientists suggests that nearly half of all plant species and more than a third of all land-living vertebrates (animals with backbones) live in just 25 small areas of the world. Some conservationists argue that if this is so, people should not worry about biodiversity everywhere. Instead, we should focus on preserving hot spots at all costs from destruction by human activity.

Large numbers of lizards are found in the biodiversity hot spots of Madagascar, Costa Rica, and the Philippines.

THE DIVERSITY OF LIFE

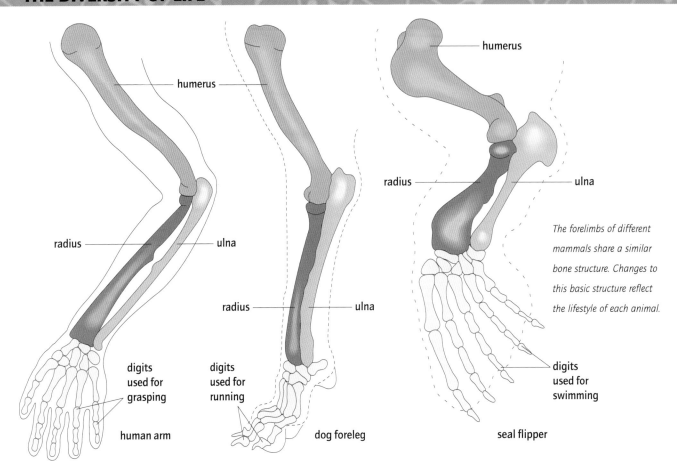

The forelimbs of different mammals share a similar bone structure. Changes to this basic structure reflect the lifestyle of each animal.

Common ancestors

The grouping of species into different levels provides information about the relationships that different species share. Since Charles Darwin (1809–1882) published his theory of natural selection in 1859, it has become clear that the groupings depend on the idea that every species is descended from a common ancestor. All mammals, for instance, descend from a single mammal species that appeared about 220 million years ago. The descendants of this animal include the first mammals with sharp cheek teeth for shearing meat. In turn, their descendants divided into species, and the order Carnivora emerged.

Among the early carnivores was a catlike animal whose descendants formed the cat family. Among them was a cat that evolved into the genus *Panthera*, which includes lions and other big cats of today.

By classifying animals and plants, scientists can construct a family tree of animal and plant species.

Looking for likenesses

The presence of shared characteristics is called homology. Biologists look, for instance, for likenesses in organisms' anatomies (the shape and structure of the body). Human arms, the front legs of a cat, whale flippers, and bat wings look very different. But under the skin the bones are remarkably similar. The basic mammal bone structure has been altered by evolution to suit each species' lifestyle, whether climbing, running, swimming, or flying.

Not all likenesses are homologies. Other likenesses, called analogies, occur when unrelated species develop similar features in response to similar environments. Bats and birds both have wings, for example. The wings look similar, but the bones supporting them are completely different. This shows that these groups developed both wings and the power of flight independently.

ECOLOGY

SEAL OF APPROVAL

Walruses, seals, and sea lions were once grouped together in the order Pinnipedia. A closer study of their anatomy recently revealed some unexpected differences. In fact, the anatomical differences were enough to suggest they may have evolved from two or more unrelated carnivores. The Pinnipedia order has now become a superfamily and part of the Carnivora order.

Information from new techniques, such as looking at organisms' genes, is making many biologists reassess traditional approaches to classification. New World vultures look very similar to those from Africa and Asia. In the 1990s, however, scientists used a technique called DNA hybridization to compare their genes. This showed that New World vultures are in fact more closely related to storks.

Sea lions are characterized by their ability to walk on all four flippers. They also lack the dense fur of other pinnipeds.

The walrus is instantly recognizable with its long, prominent tusks, whiskers, and heavy bulk. Males can weigh 4,000 pounds (1,800 kg) or more.

SCIENCE WORDS

- **carnivore** An animal that eats meat.
- **cladistics** Technique that compares large numbers of characteristics among species to build up a family tree.
- **classification** The organization of different organisms into related groups by biologists.
- **organelle** A membrane-lined structure, such as a nucleus, inside a eukaryote cell.
- **protist** A single-celled eukaryote organism with a nucleus and organelles.
- **subspecies** Subdivision of a species; a population that may have different colorings and a different range from other subspecies but can still interbreed with them.

CYCLES OF NATURE

The living world is powered by the energy of the Sun and uses basic raw materials that are constantly being recycled by natural processes.

All living organisms need energy to grow and reproduce. Nearly all of this energy comes from the Sun. The energy is absorbed by plants, which use some of it to make energy-storing foods, such as starch, in a process called photosynthesis.

Herbivorous (plant-eating) animals, such as rabbits, eat plants and use the stored energy to power their own activities. They also store some of the energy in their own bodies. If the rabbits are eaten by carnivores (meat-eaters), such as foxes, the stored energy is passed on again.

The leaves of the bright yellow sunflower harness the energy in sunlight to make food in a process called photosynthesis.

WHAT IS AN INDICATOR SPECIES?

A creature such as a bald eagle is at the top energy level of its ecosystem. An eagle needs a lot of **prey** animals to support it. The prey, in turn, need to eat a huge number of plants.

If anything interrupts the food supply lower down the ecosystem, the eagles are the first to run out of food and leave. So eagles and similar big **predators** are indicator species; that is, they show if the ecosystem is healthy.

Producers and consumers

Plants that produce food using the Sun's energy are producers. Animals, which get their energy by eating, are consumers. All consumers rely on the producers lower down the food chain. In a meadow these producers might be grass plants. There are usually many levels of consumers—the grass may be eaten by rabbits, which are then eaten by foxes, which, in turn, may be eaten by eagles.

At each level a lot of the energy is used up or lost. The total weight (or **biomass**; the total amount of living material) of grass eaten by the rabbits is not all turned into rabbit flesh. Much of it is used up by hopping around and other activities, and only a small amount is stored. In addition, animal digestion is imperfect and cannot capture all of the energy of a food item. The same happens at every level of consumption. So it takes a lot of grass to support a colony of rabbits and a lot of rabbits to support a family of foxes.

The carbon cycle

The raw materials of life are constantly recycled. One of the most important materials is carbon, the element that is the main ingredient of oil, soot, and diamonds. Carbon has a special ability to combine with other elements to make molecules of different substances, including plant and animal tissues.

ECOLOGY

TROPHIC LEVELS

Foxes and hawks both eat rabbits, but usually they do not eat each other. Ecologists say foxes and hawks are in the same **trophic level** in an ecosystem. Rabbits and mice eat grass and clover, which are on the trophic level below them. As a result of energy losses, the biomass of each ecosystem level is about one-tenth that of the level below. So, 2,200 pounds (1,000 kg) of grass and clover support 220 pounds (100 kg) of rabbits and mice but only 22 pounds (10 kg) of foxes and hawks. Thus there are many more rabbits than foxes in an ecosystem.

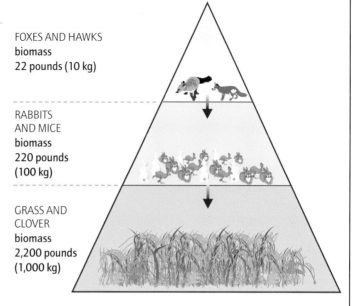

This diagram shows the trophic levels in an ecosystem that includes foxes, rabbits, and grass.

Plants get carbon from carbon dioxide gas in the air. They also take in water, which is made of hydrogen and oxygen. They use the Sun's energy to combine carbon, hydrogen, and oxygen into

Red foxes are carnivores near the top of the food chain. They eat prey such as rabbits, which are herbivores (plant-eaters).

CYCLES OF NATURE

The **carbon cycle** shows how carbon moves around Earth. Plants take carbon dioxide from the atmosphere in photosynthesis and return it during the processes of respiration and decomposition. Over many years some carbon is kept away from the cycle by dead organisms that form deposits of rocks and fossil fuels. Later it goes back into the cycle during weathering of the rocks or burning of the fossil fuels.

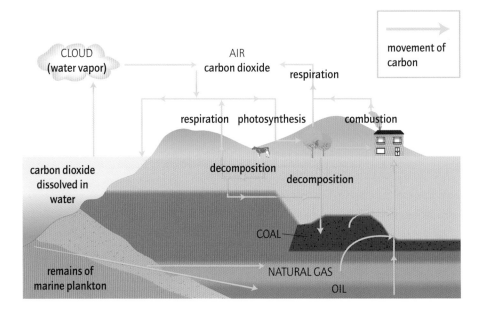

compounds called carbohydrates, substances such as sugar and starch, which store the Sun's energy. Plants use sugars to fuel cell activities or store starch. They also use sugar to make a tough plant fiber called **cellulose**, the main part of plant tissue.

TRY THIS

Making Oxygen

Water plants gather carbon from carbon dioxide dissolved in water. If you put a pondweed such as *Elodea* in a glass jar of water and put the jar in bright sunlight, you will see small bubbles forming on the leaves. The bubbles are pure oxygen, which is released as the plant changes carbon dioxide and water into sugar.

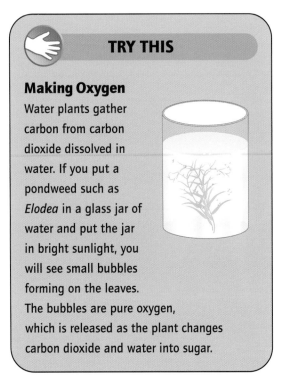

WHAT ARE FOSSIL FUELS?

When plants and animals die, the carbon in their bodies is usually recycled by other organisms and returned to the atmosphere (air around Earth). This recycling happens as the tissues decay or are broken down by microorganisms, such as bacteria. If these microorganisms cannot act on the tissues, the remains may eventually become fossilized underground as carbon-rich coal or oil. The energy locked up in these tissues is also fossilized. If these **fossil fuels** are burned, the chemical reaction releases both the stored energy and the carbon, which returns to the air as carbon dioxide.

The huge quantities of fossil fuels burned over the last 150 years have released carbon that has been stored for millions of years. This activity has increased the amount of carbon dioxide in the atmosphere. Carbon dioxide is a greenhouse gas. It is the main cause of global warming (the increase in Earth's average air temperature), which has sparked debate worldwide.

ECOLOGY

Releasing carbon

Plant-eating animals digest the sugar and starch in plants and change them into a sugar called glucose. They use it to help build their own tissues and as an energy-rich fuel. By combining the sugar with oxygen, which they get by breathing, plants change it back into carbon dioxide and water. This chemical reaction, called **respiration**, releases all the energy that the plants soaked up from the Sun. The animals breathe out the carbon dioxide into the atmosphere, where it is available to the plants again.

The nitrogen cycle

Carbon is a vital part of living tissue because it can combine with other elements to make complex molecules. One of the most important of these elements is nitrogen, which links with carbon, hydrogen, and oxygen to form proteins. Proteins are the main building blocks of animal tissue, and they are also very important to plants.

Nitrogen gas makes up 78 percent of the air, but it does not combine easily with other substances. That makes nitrogen almost useless to plants and animals in its raw state. However, in a process called **nitrogen fixation** some microorganisms turn nitrogen into compounds that are used by plants and can be turned into plant tissue. These microorganisms include bacteria that live in the soil and on the roots of plants such as clover and beans.

Most plants get the nitrogen they need from nitrates in the soil, but some can "fix" nitrogen from the atmosphere.

CYCLES OF NATURE

MANURE AND FERTILIZER

Animal dung is made up of half-digested food remains mixed with a lot of digestive juices and bacteria. Once on the ground, it decays quickly, releasing a lot of nitrogen compounds that can be taken up by plants. This action is part of the natural nitrogen cycle. Farmers can take advantage of it by spreading farmyard manure on their fields. As the manure breaks down, it adds nitrogen compounds and other nutrients, such as phosphorus and potassium, to the soil. They make the soil more fertile (rich in nutrients), so farm crops grow better. Farmers can also use fertilizer made in a factory, which contains nitrogen, phosphorus, and potassium. The chemical abbreviations for these elements are N, P, and K, so it is often called NPK fertilizer.

A farmer spreads manure from a tractor over a field. The manure is a natural fertilizer, adding valuable nutrients, such as nitrogen compounds and phosphorus, to the soil.

Decay and recycling

Plants use nitrogen to build proteins. When animals eat plants or each other, they digest the proteins and use their ingredients to make new proteins. When plants and animals die, bacteria and other microorganisms attack the proteins and break them down, releasing nitrogen compounds into the soil. The plants can absorb these types of compounds into their roots with water and recycle them to make more plant proteins.

Nitrate pollution

The factory-made fertilizers used by farmers contain nitrogen in a form called nitrate, which is ideal for use by plants. That makes it instantly effective, unlike farmyard manure, which has to be broken down to nitrate by soil bacteria. However, nitrate dissolves in water, which is how the plants absorb it. During wet weather the nitrates wash off the fields and into ponds and rivers. They then fertilize the water, causing massive growth of microscopic plantlike algae. This upsets the aquatic ecosystem, and so there is a lot of debate about the use of nitrate fertilizers.

The phosphorus cycle

Carbon and nitrogen exist in gases in the atmosphere, but other raw materials of life, such as phosphorus, potassium, calcium, and sodium, are normally in solid minerals or dissolved in water.

Phosphorus is essential to living things. It is an essential ingredient of nucleic acids, such as deoxyribonucleic acid (DNA). Plants need phosphorus to achieve healthy root growth, while animals use phosphorus and calcium to form teeth and bones. Both minerals occur in rocks. When rocks weather into mineral grains, they become part of the soil. Plants take them up in soil water and build them into their tissues. That makes phosphorus and calcium available to plant-eating animals and, in turn, to meat-eating animals. When plants and animals die, their tissues decay, so the phosphorus and other nutrients pass back into the soil.

ECOLOGY

EXTRA MINERALS

Animals need some mineral nutrients that plants do not. These minerals include sodium, an ingredient in table salt. Since most plants do not need sodium, they do not absorb it. Thus, plant-eating animals may not get any sodium in their diet. They often deal with the problem by licking the mineral from rocks or soil that contain sodium.

In Peru macaws and other birds visit clay licks to get mineral supplements. The minerals may also neutralize the effects of toxic fruits and seeds the birds eat.

Recycled rock

As rainwater drains through soil and into small streams, it carries with it nutrients, such as phosphorus, calcium, and other minerals. The streams drain into rivers that flow to the sea. Thus these minerals become available to aquatic animals, such as fish. When the fish die, their remains drift to the ocean floor, where the minerals gradually build up in layers that turn into solid rock. After millions of years the rock may be pushed to the surface and eventually form more soil.

SCIENCE WORDS

- **biomass** The total weight of all the organisms in an area (or trophic level).
- **carbon cycle** Cycle of carbon through the natural world.
- **cellulose** Tough chemical that forms part of the cell walls of plants.
- **fossil fuel** Carbon-based fuel, such as oil or coal, that forms from the remains of ancient organisms.
- **nitrogen fixation** The incorporation by soil bacteria of nitrogen in the air into nitrate compounds that plants are able to use.
- **predator** Animal that catches other animals for food.
- **prey** Animal caught and eaten by another animal.
- **respiration** The process of reacting food with oxygen to liberate energy; takes place inside the mitochondria of eukaryotes or on cell membrane of prokaryotes like bacteria.
- **trophic level** One of the links in a food chain; plants, herbivores, and carnivores are all at different trophic levels.

The colorful blue-and-gold macaw is native to the tropical woodland of South America. These birds get extra minerals by visiting clay licks.

21

CLIMATE AND EARTH

Each place on Earth has its own climate, which is that region's regular pattern of weather. Any factors that affect a region's climate will have an effect on the types of living organisms that thrive there.

Climate is a broad picture of the average weather conditions experienced in a region over many years. The Sun provides the energy that drives the weather and so produces climate. It causes winds to blow and moisture to circulate between the atmosphere and the surface of Earth in a continual

Palm trees bend under the force of powerful winds called hurricanes, which blow at more than 75 miles (120 km) per hour. Hurricanes start above ocean areas in the world's tropical regions.

process called the water cycle. All the day-to-day weather conditions, such as sunshine, showers, storms, and blizzards, are caused by the Sun heating various parts of Earth by different amounts. Differences in temperature produce variations in air pressure, the force exerted on Earth's surface by the weight of the air around it. Changes in air pressure then cause air masses to move from one place to another, which produces winds. Winds may bring warm or cool temperatures and dry or rainy spells.

CLIMATE ZONES

Austrian scientist Wladimir Köppen (1846–1940) was among the first to classify regions according to climate. Köppen studied how plant life in different parts of the world is affected by climate. In 1918, he produced a classification listing five main climate zones: tropical/rainy, dry, temperate/rainy, cold/snowy, and polar. Köppen's classification is still used, but many modern climatologists (climate experts) distinguish at least a dozen climate zones, as shown on the map below.

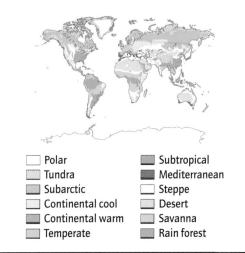

- Polar
- Tundra
- Subarctic
- Continental cool
- Continental warm
- Temperate
- Subtropical
- Mediterranean
- Steppe
- Desert
- Savanna
- Rain forest

ECOLOGY

SUN'S POWER

The Sun's rays travel in parallel lines toward Earth. They strike down from directly above at the equator, where they are concentrated on a relatively small area. So the climate in the tropics is always hot. The Sun beats down less directly in temperate regions, which are north and south of the tropics, so they have a cooler climate. At the polar regions, where Earth's surface is most curved, the rays strike at a sharp (oblique) angle and are spread out over a much wider area. The rays also have farther to travel through the atmosphere, which absorbs heat, so the polar climate is cold.

The Sun's rays hit Earth's surface at different angles.

Factors affecting climate

The three main factors influencing climate are latitude (distance from the equator), altitude (height above sea level), and how far a region is from the sea. Latitude is a very important factor. Earth's curving surface enables various regions to receive differing amounts of the Sun's heat.

The seasons

Many parts of the world experience regular seasonal changes because Earth tilts on its axis (an imaginary line linking the North and South poles) at an angle of 23.5 degrees as it orbits the Sun. At any time one hemisphere (half of Earth) leans toward the Sun and has summer with long hours of daylight. The other hemisphere tilts away and has winter with short days.

The tilting of Earth on its axis as it goes around the Sun means that the northern hemisphere is tipped closer to the Sun in June, making longer and warmer days in the season called summer. In June the southern hemisphere is tipped farther from the Sun, causing winter, a season with

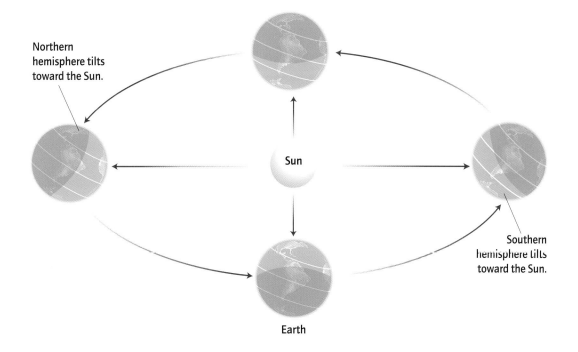

Northern hemisphere tilts toward the Sun.

Sun

Earth

Southern hemisphere tilts toward the Sun.

23

CLIMATE AND EARTH

colder and shorter days. The opposite happens in December.

Seasonal changes are least noticeable in the tropics, the regions closest to the equator that always face the Sun. However, some parts of the tropics have a dry and a rainy season. Many temperate regions have four seasons: spring, summer, fall, and winter. The poles have the greatest variation in temperature and day length. For part of the summer each pole is bathed in weak sunlight for 24 hours a day. For part of the winter the Sun never rises. Arctic plants cope with this by becoming dormant (inactive) in winter. Many animals hibernate (enter a sleeplike state) or migrate (move away) to avoid the cold.

Kilimanjaro, the tallest mountain in Africa, is 19,340 feet (5,895 m) at its highest point. It is just a few miles from the equator, but it is always capped by snow because the thinner air at high altitude absorbs less of the Sun's heat.

TRY THIS

Recording Seasonal Changes

Monitor how seasonal changes affect the area where you live by keeping records. Record temperatures at regular times each day for one week each month, and note the hours of daylight. You could also record how the seasons affect local plants and animal behavior. For example, local mammals and reptiles may not be seen in winter because they hibernate, and some local birds may migrate.

ECOLOGY

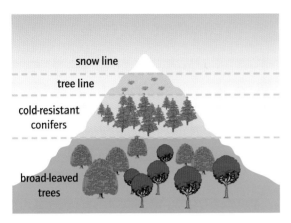

This diagram shows the different climate zones on a mountain. Only a few plants can survive above the tree line, and almost no plants are found above the snow line.

Altitude

Height above sea level, or altitude, also affects a region's climate because the thinner air at higher altitudes absorbs less of the Sun's heat. For every 500 feet (150 m) climbed, the temperature drops about –1.8°F (1°C). Mountain climates are thus much cooler than surrounding lowlands, with shorter summers and longer winters. The tops of high mountains near the equator, such as Kilimanjaro in East Africa, are always covered with snow.

Altitude affects the types of plants that grow at different heights on mountains. Generally, higher up on mountains only small plants grow. For example, broad-leaved trees, such as oaks, may grow at a mountain's base, but only cold-resistant conifers, such as pine trees, thrive higher up. Even higher, beyond a zone called the tree line, there are no trees, only low-growing plants. The kinds of plants that can grow at that altitude are usually found in alpine **tundra** regions. Beyond the snow line almost no vegetation grows because the climate is too harsh.

Maritime and continental climates

Distance from the sea also affects a region's climate. The oceans absorb the Sun's heat more slowly than the land but retain their warmth longer. Coasts may also be warmed or cooled by ocean currents and are thus usually milder than regions far inland and also wetter because winds blowing inshore off the sea are laden with moisture. Coastal regions have a maritime climate, while areas far inland have a continental climate.

The atmosphere

The atmosphere is a blanket of gases surrounding Earth like the peel on an orange. It extends about 450 miles (720 km) into space. The main gases in the atmosphere are nitrogen (78 percent) and oxygen (21 percent). Smaller amounts of carbon dioxide, water vapor, nitrous oxide, ozone, and other gases are also present. They are important because they trap the Sun's warmth. The atmosphere shields Earth from radiation, keeps out the cold of space, and contains oxygen for living things to breathe. Without it life on Earth would be impossible.

About 75 percent of all the air in the atmosphere exists below 6½ miles (10.5 km). The air becomes progressively thinner closer to space. Scientists distinguish five main layers in the atmosphere. The lowest layer, the **troposphere**, extends to 12 miles (19 km) above the planet's surface. Most weather occurs here. Above that is a calm layer, the **stratosphere**, extending to about 30 miles

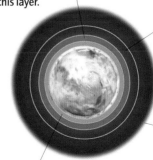

Jet planes fly in the **stratosphere**. The ozone layer is in the upper part of this layer.

The **mesosphere** has the same mixture of oxygen, nitrogen, and carbon dioxide as lower layers, but it has very little water vapor.

The **troposphere** contains three-quarters of the water vapor in the atmosphere. Nearly all the clouds, rain, and snow occur in this layer.

Spectacular light effects called auroras can be seen in the **thermosphere**.

The different layers of the atmosphere are the troposphere, stratosphere, mesosphere, thermosphere, and exosphere.

25

CLIMATE AND EARTH

(48 km) up. Aircraft fly here to avoid turbulence. Most meteors burn up in the layer beyond, the **mesosphere**, 30 to 50 miles (48–80 km) up. The thermosphere—50 to 280 miles (80–450 km) up—is a layer containing electrically charged particles. Beyond that a layer of very thin air called the exosphere extends into space.

The ozone layer

A layer of ozone gas is present in the stratosphere about 15 miles (24 km) above Earth's surface. Ozone, a form of oxygen, is a bluish gas. The **ozone layer** shields Earth from harmful ultraviolet (UV) radiation in sunlight. UV rays can cause illnesses such as cancer and eye damage in animals and also harm plants and **phytoplankton**, a major food source in the sea.

The greenhouse effect

Carbon dioxide, methane, water vapor, and other gases in the atmosphere act as a barrier that traps heat rising from Earth's surface and prevents it from escaping into space. These gases have a similar effect as the glass in a greenhouse, so this process is called the **greenhouse effect**. For millions of years

OZONE LOSS

In the 1980s scientists discovered that a hole in the atmosphere containing less ozone than normal was appearing over Antarctica each spring. During the 1990s ozone holes over Antarctica and the Arctic got steadily bigger. Scientists showed that chlorofluorocarbons (**CFCs**) were mainly causing the damage. CFCs are chemicals used in the manufacture of refrigerators, polystyrene packaging, and aerosol spray cans, and in air conditioning. The loss of ozone has caused an increase in skin cancer in people and animals.

CHANGING SEA LEVELS

The diagrams BELOW show (a) the position of Florida's coastline 18,000 years ago, when the sea level was lower than present, and (b) and (c) the possible situation if global warming causes the polar ice cap to melt. Most of Florida would disappear completely.

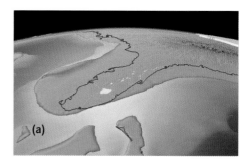

This illustration shows the sea level as it was 18,000 years ago—394 feet (120 m) lower than at present.

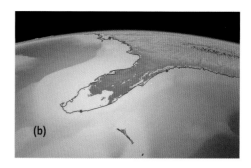

This illustration shows the future sea level 17 feet (5 m) higher than at present.

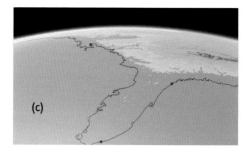

This illustration shows the possible future sea level 170 feet (50 m) higher than at present.

ECOLOGY

the greenhouse effect has created a warm environment on Earth that has allowed living organisms to flourish here. Scientists now fear that human-made pollution is increasing Earth's greenhouse effect. Earth's climate has been growing steadily warmer in a process called global warming. Temperatures rose by 1°F in the last 50 years and may now be rising more rapidly. Scientists have found that human-made pollution is causing the warming. As people burn fossil fuels in cars, power plants, homes, and factories, more carbon dioxide is added to the atmosphere, increasing global warming. This is resulting in the melting of the polar ice, causing the sea levels to rise and flooding of coastal land.

In some African countries, people have to walk many miles to get the water they need each day.

Climate change

Earth's climate appears to stay the same, but it changes very slowly. Long, cold periods called ice ages gradually give way to warmer ones. Scientists believe these changes are caused because Earth wobbles slightly as it orbits the Sun. During the past 2 million years a total of 15 ice ages have come and gone. The last major ice age ended about 10,000 years ago. During these cold periods ice covered large parts of what are now temperate regions, including much of North America and Europe.

 TRY THIS

Saving Water

The world's rainfall is not evenly distributed over the planet's surface. Some regions receive high rainfall, while in deserts rain may not fall for years. Where water is scarce, people may have to spend hours every day just fetching enough to meet their needs. Elsewhere water may be plentiful, but purifying and distributing it still use up energy and cost money. Do you think it is important to save water? If so, can you think how it can be done?

CLIMATE AND EARTH

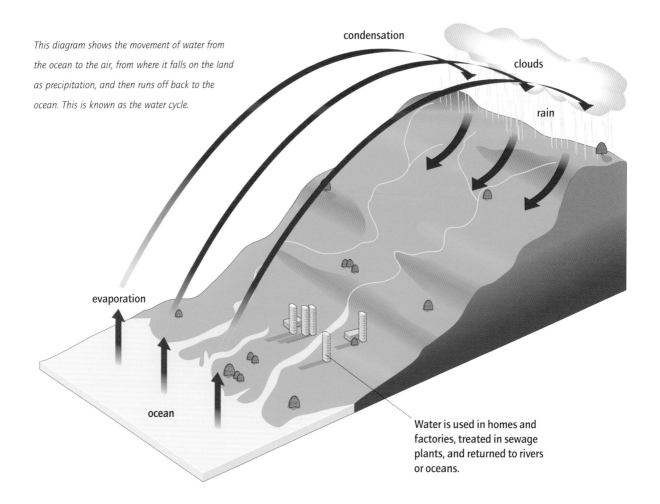

This diagram shows the movement of water from the ocean to the air, from where it falls on the land as precipitation, and then runs off back to the ocean. This is known as the water cycle.

Water is used in homes and factories, treated in sewage plants, and returned to rivers or oceans.

Scientists find out about Earth's climate in the past using various techniques. In one method, they collect samples of ice buried deep in the polar regions and examine them to find out what the climate was like many thousands of years ago. Studying tree rings also yields climate information.

Life-giving water

As well as oxygen from the atmosphere, all organisms need water, which makes up on average 75 percent of living things. Water exists on Earth in three forms—as a liquid, as a gas (water vapor), and in solid form as ice. Moisture collects in the atmosphere to form clouds made up of millions of tiny water droplets. Clouds shed moisture as precipitation in the form of rain, hail, or snow.

The water cycle

Moisture moves constantly between the air, land, and oceans in a process called the water cycle. The Sun's heat causes moisture from oceans, lakes, and wet ground to evaporate (turn into water vapor). At cooler temperatures the moisture condenses (turns into liquid) to form clouds, which later shed rain or snow. When rain falls, the excess moisture not absorbed by the soil or taken up by plants drains into streams, rivers, and lakes, which eventually empty into the ocean, and so the cycle goes around again.

Soil and climate

Soil forms the link between the living world and the nonliving rocks that make up Earth. Most living things on land need the soil either directly or indirectly to

ECOLOGY

live. Plants root in the soil and take water and nourishment from it. An area of soil just 1 foot (30 cm) square may contain literally millions of organisms, including insects, spiders, slugs, and worms, and microscopic bacteria and fungi. Many of them help break down plant and animal remains so their nutrients return to fertilize the soil.

Over thousands of years the forces of weather slowly create soil, as ice, frost, wind, and running water break down the rocks at Earth's surface in a process called erosion. Different types of soil, including chalk, clay, sand, and peat, are present in different regions. The type of soil depends on many factors, including climate, vegetation, and the composition of the underlying rock.

SCIENCE WORDS

- **greenhouse effect** The rise in global temperatures caused by an increase in gases such as methane and carbon dioxide in the atmosphere.
- **mesosphere** Atmospheric level between troposphere and stratosphere containing little water vapor, and where the temperature drops to −112°F (−80°C).
- **ozone layer** Layer of ozone gas high in the atmosphere. It filters out harmful ultraviolet radiation from the Sun.
- **stratosphere** Upper level of the atmosphere. It contains the ozone layer; extends from the mesosphere to the edge of interplanetary space.
- **troposphere** Lowest layer of the atmosphere. It contains most of the water vapor; clouds occur in this layer, and weather changes happen there.

Soil profiles

Scientists distinguish several different layers called horizons in a vertical slice of soil, called a soil profile, see left. The depth of each layer varies with the type of soil. The topmost layer (horizon O) is a rich layer of humus made of plant and animal remains. Below that the topsoil (horizon A) is a dark, fertile layer containing humus. The subsoil (horizon B) is nourished by minerals washed from the topsoil. The zone of partly weathered rock below that (horizon C) is less fertile. Last comes the bedrock (horizon D or R), the source of the minerals in the soil.

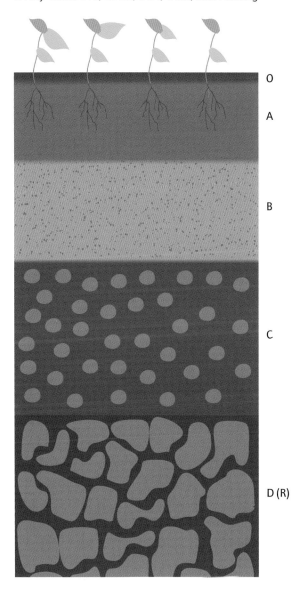

This diagram shows the different layers, or horizons (O, A, B—see text right), through a slice of soil. This is called a soil profile.

29

ECOSYSTEMS

Ecosystems are self-contained units made up of living things and their nonliving environment. An ecosystem can be as small as a tiny pool in a dry canyon or as vast as the ocean.

Luckily for people and all Earth's organisms, a number of factors come together to make our planet suitable for life. First, Earth is just the right distance from the Sun to receive the light and heat needed for

This colorful alpine meadow is just one example of an ecosystem. It supports a wide variety of animals, plants, and microorganisms.

THE BIOSPHERE

The **biosphere** is all parts of Earth that are able to sustain life, including the atmosphere, oceans, the surface of the land, and just below it. This living layer extends from about 30,000 feet (9 km) in the air down to the ocean depths and to about 8 miles (13 km) below the land's surface. Above 30,000 feet the air is too thin to support life, while at depths below 8 miles (13 km) the heat and the pressure are too great. Within this living layer most life exists in a narrower band, which extends from the upper waters of the oceans to about 1,000 feet (300 m) in the air.

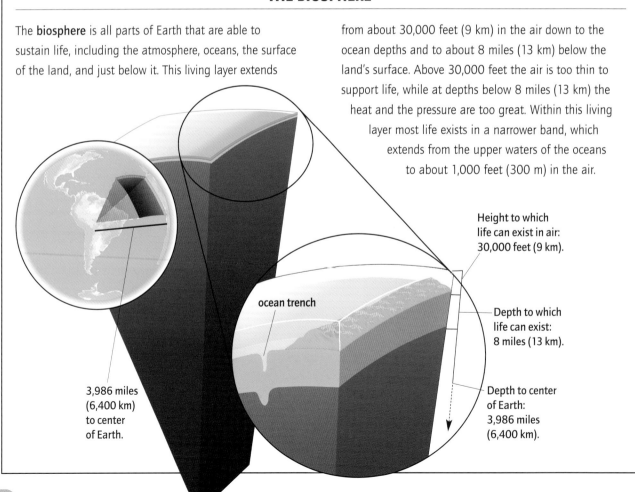

3,986 miles (6,400 km) to center of Earth.

ocean trench

Height to which life can exist in air: 30,000 feet (9 km).

Depth to which life can exist: 8 miles (13 km).

Depth to center of Earth: 3,986 miles (6,400 km).

ECOLOGY

life to flourish. Earth's atmosphere protects us from harmful rays from space and contains life-giving oxygen. There is plenty of water, and the oceans as well as the atmosphere help regulate temperatures. In short, Earth has all the conditions needed to support life except an energy source, and that is provided by our local star, the Sun.

In contrast, Mars is too far from the Sun to sustain life. The average temperature there, –16°F (–23°C), is too cold for life, since the atmosphere is too thin to retain heat. Venus, closer to the Sun than we are, is too hot for life, with average temperatures of 480°F (234°C). Its thick, dense atmosphere has clouds of carbon dioxide. Also, neither of the planets Mars and Venus has enough water to support life.

BIOMES

Earth is divided into a number of huge ecosystems such as deserts, forests, tundra, and mountains. They fit together like the pieces of an enormous jigsaw puzzle. Each piece is a biome, with its own climate and distinctive soil, plants, and animals. The climate, geography, and other factors allow certain types of vegetation to grow well. This in turn supports particular types of animals, fungi, and microscopic life.

Some biomes support vast numbers of living organisms, while other biomes are much less productive. Of the dry land that occupies about 30 percent of Earth's surface, one-third is desert, and another fifth is either tundra (treeless areas where the subsoil is frozen) or permanently covered with ice. These harsh biomes support relatively little life. Biomes such as forests, grasslands, and warm, shallow seas teem with life.

The polar regions

The polar regions are the harshest places on Earth, with short, cold summers and long, dark, freezing winters. A desert is anywhere with less than 10 inches (25 cm) of rain every year, so polar regions are deserts because there is very little rainfall and

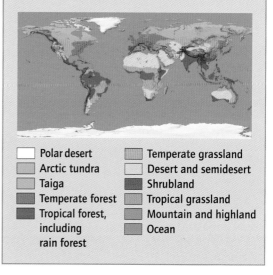

WORLD'S BIOMES

Biomes are large regions that have similar climate, plants, and animals. This map shows Earth's main land biomes: polar, tundra, mountain, desert, scrubland, grasslands, and forest. Forest biomes can be divided into cool, temperate, and tropical forests. Oceans and lakes offer saltwater and freshwater biomes. Each biome contains ecosystems with particular conditions. The distance from the equator is one factor that determines climate, so divisions between biomes broadly follow lines of latitude.

- Polar desert
- Arctic tundra
- Taiga
- Temperate forest
- Tropical forest, including rain forest
- Temperate grassland
- Desert and semidesert
- Shrubland
- Tropical grassland
- Mountain and highland
- Ocean

almost no liquid water on the ground. A thick cap of ice, more than 2 miles (3.2 km) deep in places, permanently covers the land. The oceans are warmer but still frozen over for much of the year. Little life can survive on land in these bleak conditions; but the polar oceans, with their nutrient-rich currents, teem with life, from microscopic plankton to Earth's largest creatures, giant blue whales.

The tundra

The bleak, treeless lands that lie south of the arctic wastes are the tundra. There the climate is also harsh, with cool, brief summers and long, icy winters. Snow and ice cover the ground for much of the year,

ECOSYSTEMS

The polar regions surrounding the North and South poles are actually examples of desert biomes because so little water falls as precipitation (rain or snow).

but in summer the snow melts to reveal grassy lowlands with bogs and lakes. Beneath the topsoil is a layer of permanently frozen ground called the permafrost in which trees cannot root. However, low-growing shrubs, mosses, lichens, and flowering plants make the most of the short summer growing season. Few animals live all year on the tundra. Insects die in the fall, leaving their eggs or larvae (young) in the soil to emerge the following spring. Birds and mammals that live year round include arctic foxes, lemmings, and snowy owls. Many more, including caribou, geese, ducks, and seabirds, migrate to the tundra to breed in spring.

Mountain biomes

The effects of altitude on climate create mountain zones that contain several minibiomes with conditions similar to those normally encountered over a huge north–south range. The summits of high mountains are permanently covered with snow and ice, like the polar regions. Below that is a bleak, treeless zone with vegetation similar to the tundra. Farther down, lower slopes may feature cool forest and then temperate or tropical plant life, depending on where the mountain is located.

Grasslands

Grasslands cover roughly 13 million square miles (34 sq km) worldwide—about one-quarter of Earth's land area. Most grasslands lie between desert biomes and tropical or temperate forests. There are two main types of grasslands: **savannas**, or tropical grasslands, and temperate grasslands. Tropical grasslands generally experience an annual dry and rainy season, and have scattered trees and shrubs. Temperate grasslands receive a more even distribution of yearly rainfall.

The major temperate grasslands of the world include the central Asian steppes and the American prairies and pampas. Huge tracts of these vast grasslands have been plowed to grow crops such as wheat, corn, and oats. Wild grasslands of the world are home to 7,500 different species of grasses but relatively few trees. Grasslands are rich biomes that support huge numbers of insects and microorganisms, and a variety of birds and reptiles. Grassland mammals fall in two main groups: burrowers and grazers. Burrowers include ground squirrels and rabbits, while grazers include antelope, zebras, and wildebeest.

Forest biomes

Forest biomes together cover over one-quarter of Earth's land area. There are three main types of forests:

DUST BOWL

In the 1930s, bad farming practices turned a huge area of the American prairies into a barren wasteland. As the West was settled, farmers removed the natural vegetation to plant crops. Without grass roots to anchor the soil, year after year of drought reduced it to dust. High winds blew the dust away, creating a desert often called a dust bowl. The region's fertility was restored only after many years of careful management.

ECOLOGY

The tropical grassland of central Africa is known as the savanna. Predatory lions hunt herd animals such as wildebeest and zebra on these vast, open plains.

coniferous, deciduous, and tropical forest. Climate is the main factor that determines which type of forest grows.

A wide belt of coniferous forest called the taiga rings the northern hemisphere south of the tundra. Conifers such as spruce, larch, fir, and pine thrive in the harsh climate there, with bitterly cold winters and low rainfall. These cone-bearing trees have narrow, waxy leaves called needles that minimize moisture loss and withstand cold well. All except larch are evergreen. The dense foliage throughout the year severely limits the amounts of both light and moisture that filter through to ground level, so plant and animal life is relatively sparse on the forest floor.

Warm summers, cool winters, and abundant rainfall generally characterize the world's temperate regions. Here forests of deciduous trees such as oak, beech, ash, and maple shed their leaves in fall to conserve moisture in winter, and then grow new

TRY THIS

Make a Mini Rain Forest

Make a mini rain forest using an old fish tank or a large jar with a lid. Cover the base of the container with gravel, and then add charcoal and a layer of compost. Plant tropical plants from a garden center in the soil, mist their leaves with a plant spray, and then replace the lid. The plants will recycle the moisture inside their mini-ecosystem, so you will not need to water them much.

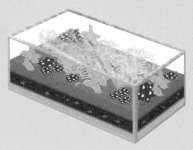

33

ECOSYSTEMS

leaves in spring. For much of the year the bare branches allow abundant light and moisture to reach the ground, where wildlife thrives in the rich soil. In tropical areas with an annual dry and rainy season trees also shed their leaves in the dry season.

Tropical rain forests grow in low-lying regions near the equator that receive more than 100 inches (250 cm) of rain a year. Vegetation thrives in these warm, wet conditions and supports an abundance of wildlife. Rain forests cover just 6 percent of Earth's land area, yet scientists estimate they host up to 70 percent of all living species. The world's largest rain forest, the Amazon rain forest in South America, occupies a vast area. However tropical forests worldwide are disappearing at an alarming rate.

This aerial photograph shows part of the Amazon rain forest in South America. The Amazon has unparalleled biodiversity—one in ten of the world's known species live there.

Desert and scrubland biomes

Deserts are defined as places where less than 10 inches (25 cm) of rain falls annually. Scrublands, often on the borders of deserts, are slightly wetter but still have a low rainfall. In low-lying deserts, such as the Sahara in Africa, temperatures may climb to 130°F (54°C) by day but drop steeply at night. High deserts such as the Gobi in central Asia have an even harsher climate, with temperatures that go below freezing at night.

Desert organisms have features that enable them to cope with the harsh conditions. Many desert plants have a deep or extensive root system to gather moisture from the soil. Plants called succulents, such as cacti, store moisture in their fleshy leaves, stems, or roots. Many animals are active by night, when it is cooler, and are able to withstand a period of drought.

EVOLUTION ON ISLANDS

English naturalist Charles Darwin (1809–1882) visited the remote Galápagos Islands off Ecuador in the 1830s. His experiences there helped him formulate his ideas on evolution. He discovered many of the islands had their own species of birds called finches, with beaks specialized for different foods. Darwin speculated that all had evolved from a single species of finch that had reached the islands several hundred thousand years before.

ECOLOGY

Deserts may seem barren and devoid of life, but a few hardy animals and plants can survive in the extreme heat of the day—and the extreme cold at night.

Urban ecosystems

As human populations grow and expand around the world, so more and more wild land is transformed into urban environments. For many species this change is disastrous, but some types of plants and animals have adapted to thrive in this new habitat. Opportunistic feeders such as raccoons and foxes were once woodland species but now also survive in cities. Birds such as pigeons do well in cities too because they are able to exploit urban food sources.

Islands

Islands often have very different communities of plants and animals than the nearest mainland. The isolation of island habitats ensures that only certain species of plants and animals are able to reach and colonize them. Plants whose seeds are dispersed by wind and water, and flying animals such as birds, bats, and insects are common, but large, land-based predators are often absent. These conditions may cause island species to evolve along different lines from their mainland cousins. For example, many birds that have evolved on islands, such as New Zealand's kiwis, became flightless in the absence of predators. Island species are especially vulnerable to changes in their environment, whether these changes occur naturally or if humans cause them.

Population ecology

In the wild, animal populations are held in balance by natural checks such as disease. In a particular year animal numbers may rise or fall, but they usually return to a consistent level over the course of several years. Competition for food among animals of the same species is one important factor. Different species within a community do not compete directly for food because they have different niches and eat slightly different foods.

ECOSYSTEMS

Predator–prey relationships

In any ecosystem predators curb numbers of **herbivores** (plant-eaters), and that in turn protects the plants at the base of the food chain. Many animal populations undergo a regular cycle of "boom and bust," which relates to factors such as climate. In years of harsh weather plant food is scarce, so herbivore and carnivore numbers remain low. In mild years plants thrive, and herbivores breed quickly. Increased numbers of prey prompt the predators to multiply; but when they become too numerous, populations of their prey may crash. Some predators starve, so their numbers drop, which allows the herbivores to recover. In this way balance is restored.

Territories

Many different types of animals, from fish to birds and mammals, establish group or individual territories. Each animal's "patch" contains enough food for the individual or group's needs, or is a safe place where the animals can breed or shelter. Animals defend their patch against others of their

Many birds, such as these gannets, gather in vast nesting colonies, with clearly defined territories within the nesting site.

POPULATION EXPLOSION

In contrast to animal populations, which do not usually keep rising, the number of people on Earth has risen steeply in the last few centuries. Improved health care and more efficient farming practices mean that each year, more and more people live to an age at which they themselves can have children, so the human population "explodes." This huge increase puts great pressure on the natural resources on Earth and on ecosystems worldwide.

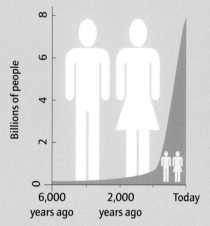

This diagram shows how the human population has exploded in the last few millennia.

ECOLOGY

A lion's territory is determined by the size of the pride and the availability of prey and water. All the lions in the pride will defend their territory from other lions.

species, which limits the number of any one species a habitat will support. Animal territories thus help keep populations in balance.

Animal **territories** vary greatly in size. For example, a pride of lions needs a huge territory in which to hunt, while seabirds such as common murres build tiny territories on narrow cliff ledges just large enough for each female to lay eggs.

SCIENCE WORDS

- **biosphere** All the parts of Earth that are able to sustain life; includes the lower atmosphere, oceans, and the surface of the land and fresh waters, and extends to a mile or so below the surface.
- **herbivore** Animal that feeds on plants.
- **savanna** A tropical grassland.
- **territory** Area controlled by an organism to protect food supplies or to attract a mate.

37

WATERY WORLD

Water covers 71 percent of Earth's surface. Aquatic ecosystems can be divided into freshwater systems, such as lakes, rivers, and wetlands, and saltwater systems: seas and oceans. Semisalty, or **brackish**, waters are present in **deltas**, estuaries, and marshes, where freshwater mixes with the tide.

Four main factors affect the nature of aquatic ecosystems. They are **salinity** (salt content), oxygen levels, the amount of sunlight that reaches the water, and water temperature. Salinity is measured in parts per thousand (ppt). Seawater contains 35 to 70 ppt, freshwater only 15 to 30 ppt.

Sunlight penetrates only the upper waters of aquatic biomes. Moving waters such as swift-flowing streams and the surface of rough seas are particularly high in oxygen. The upper waters are also warmer than the depths, and water near the surface contains the most oxygen.

> **WHAT MAKES SEAWATER SALTY?**
>
> Seawater contains dissolved minerals from the land, which are washed into seas by rivers. The main minerals are sodium and chlorine in the form of salt (sodium chloride). Seawater also contains smaller amounts of other minerals, including calcium, potassium, and magnesium.

Marine ecosystems

The oceans are by far the largest biome on Earth. They occupy 70 percent of the planet's surface and support an estimated 250,000 species of living organisms. Marine organisms can be divided into three groups, of which one is plankton—generally, small creatures and microorganisms that float on the water. The other groups are larger, actively swimming creatures, including fish, squid, and marine mammals; and smaller, less active creatures, including starfish, corals, sea anemones, sea lilies, and sponges that live on the bottom of the ocean.

Marine ecosystems can be divided into two main biomes: the deep ocean and the tidal zone—the shallow waters that edge the coasts. Tidal zone biomes include coral reefs, which are sometimes

The world's oceans cover more than 70 percent of Earth's surface, making it the largest biome on Earth.

38

ECOLOGY

compared to rain forests because they support a huge biodiversity. In addition, there are shoreline ecosystems and semisalty habitats such as deltas and estuaries.

The deep, open ocean can be divided into vertical zones that can be compared to the vertical layers called stories that divide forest ecosystems. There are three main zones: the sunlit zone, the midwater, and the deep. Life is present at all depths of the ocean, but the sunlit upper waters are very productive and have the most biomass (amount of living material).

Upper waters

Plants and algae, including seaweeds, grow only in the sunlit upper waters of the ocean or in coastal shallows on the seabed. In spring the phytoplankton multiply, or "bloom," in response to

Moon jellyfish glow yellow in the blue water of the open ocean.

AQUATIC FOOD WEBS

Almost all life in aquatic habitats depends on the Sun for energy, as on land. Aquatic food webs are as vast and complex as land-based webs. At the base of the food chain tiny plantlike, single-celled algae called phytoplankton floating at the surface harness the Sun's energy through photosynthesis. In so doing, the phytoplankton produce a surprising 70 percent of the world's oxygen. Tiny animals called **zooplankton** feed on the phytoplankton and, in turn, provide food for larger creatures, and so on up the food chain. In the oceans small crustaceans called copepods and krill, part of the zooplankton, are food for the largest creatures, baleen whales. When aquatic organisms die, their remains are eaten either by scavengers such as shrimp and crabs or by microorganisms. The microorganisms break down organic remains into simpler compounds so their energy can be recycled.

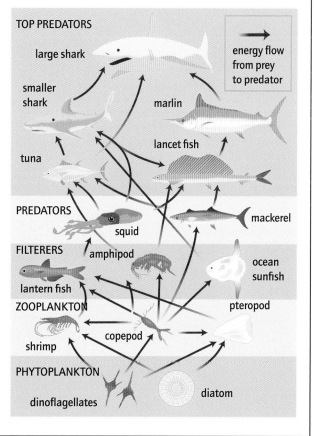

This diagram shows some of the species that make up an aquatic food web.

WATERY WORLD

Seaweed litters the tideline at low tide. It is an important source of food for other marine organisms.

the longer hours of daylight, creating an abundance of food for herbivores (plant-eaters). Some free-swimming creatures, including filter-feeding baleen whales, migrate long distances to feast on phytoplankton—or on herbivores.

Many fish of the upper waters, such as mackerel and herring, have dark backs and pale bellies. This coloration, termed **countershading**, helps conceal them both from seabirds hunting in the air above and from predators below. The seabirds cannot easily see the fish while looking toward the darkness of the water from above, nor can the ocean predators see the fish while looking toward the light from below.

Bottom dwellers

The ocean floor is as varied as the surface of the land, with features such as high ridges, tall peaks called seamounts, vast plains, and deep trenches. Bottoms may be rocky, sandy, or in the case of many deep oceans, covered with a layer of silty ooze. Plaice, rays, and other bottom-dwelling fish are adapted to their habitat with wide, flattened bodies for moving along the seabed. Some deep-sea dwelling creatures, such as sea spiders

> ### OCEAN ZONES
>
> The sunlit upper waters, called the **euphotic zone**, extend to a depth of 650 feet (195 m). Beyond that lies the midwater zone, which extends to about 6,500 feet (1,950 m). The upper waters of the midwater are sometimes called the **twilight zone** since some light reaches here, but no light penetrates below 3,000 feet (900 m) to the region called the **dark zone**. The deep-sea zone lies below 6,500 feet and may include ocean trenches descending to 35,000 feet (10,500 m). Each zone is home to a distinctive community of organisms; however, deep-diving creatures such as sperm whales regularly move between the zones.
>
>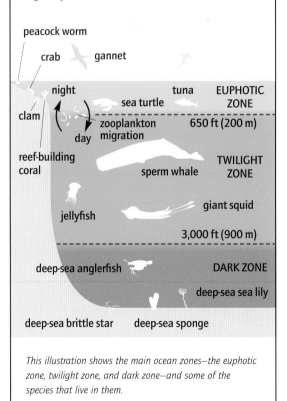
>
> *This illustration shows the main ocean zones—the euphotic zone, twilight zone, and dark zone—and some of the species that live in them.*

and the extraordinary tripod fish, have long legs or leglike spines that act as stilts and hold the animals high above the ooze on the ocean floor.

ECOLOGY

Wildlife of the depths

Food is scarce in the dark ocean depths. Fish and other deep-sea creatures feed on plant and animal remains that drift down from the waters above or prey on one another. Fish such as gulper eels and viperfish have huge jaws, sharp teeth, and stretchy stomachs so they can make the most of any prey they come across. Some deep-sea animals can swallow creatures twice their own size.

In the twilight zone, where dim light penetrates, some fish have rows of little lights on their bellies to disguise their shadow. The light may be produced either by special organs called photophores or by luminous (light-emitting) bacteria in the fish's skin.

In the deep oceans scientists discovered clams and tubeworms living around **hydrothermal vents**. These vents are cracks or chimneys from which black, billowing clouds of superhot, sulfur-rich water well up from inside Earth's crust.

The clams and worms do not depend on sunlight to make their food but get all their nutrients from bacteria inside their bodies. In turn, the bacteria get energy to make food from the sulfur present in the hydrothermal vents.

Coastal and intertidal zones

These zones include coral reefs and seashores. Coral reefs are present mostly in warm, shallow waters off tropical coasts. The reefs are made of the chalky skeletons of millions of small, sea anemone-like reef-building corals called coral polyps that build up on top of one another over thousands of years. Some coral reefs are 40–50 million years old and are made up of coral three-quarters of a mile (1.2 km) thick. The world's largest coral reef, Australia's Great Barrier Reef, stretches over 1,240 miles (2,000 km) and can be seen from space. In places it towers 500 feet (150 m) above the seabed. Coral reefs occupy only a small fraction of coastal waters worldwide, but these ecosystems support an estimated 25 percent of marine life, providing food, shelter, and hundreds of hiding places for many creatures. For instance, a reef at Key Largo, Florida, has more than 550 species.

Coral reefs are the ocean's richest habitats, but these amazing structures are now under threat throughout warm oceans. Pollution is one of the main dangers.

WATERY WORLD

TRY THIS

Exploring Pool Life

Each pool on the beach is a mini-ecosystem. What can you find out about life in a pool on the nearest shore? First, is the beach sandy, rocky, or shingly? Is the pool situated on the upper, middle, or lower shore? Now analyze the pool itself. Is it deep? Is the water very salty, or does it contain rainwater? Can you see any plants or seaweed? Now try to spot herbivores such as limpets and periwinkles, predators such as dog whelks, and scavengers such as shrimp and crabs. Can you draw a food web connecting the organisms you see?

sea urchin

shrimp

seaweed

Seashores

Worldwide, there are more than over 300,000 miles (480,000 km) of coastline. Rocky, shingly, sandy, and muddy beaches each host their own communities of life. Rocky shores are very rich in wildlife, including seaweeds, barnacles, shrimp, limpets, sea anemones, and crabs. Cockles, clams, and razor shells live on sandy beaches. Many sandy-beach animals burrow to hide from predators and to stay moist when the tide is out.

Seashores are harsh habitats, where conditions are continuously changing. The animals and plants living there spend part of each day covered by water and the remainder exposed to the weather, which may be baking sunshine, winds, or rain. Organisms have to adapt to these two different lifestyles. They time their resting and feeding periods to match the rhythm of the tides.

Rock pools support many different marine organisms, such as crabs, sea anemones, kelp, limpets, sea urchins, and some types of fish.

ECOLOGY

SEASHORE ZONES

Seashores are divided into three zones between the high- and low-tide mark: the upper, middle, and lower shore. Each is home to distinctive plants, animals, and other life. Beyond the upper zone lies the splash zone, where rocks are misted by salt spray and stained with lichen. Limpets, periwinkles, barnacles, shrimp, and wrack seaweeds are present on the upper and middle shore. Lobsters, sea anemones, sea urchins, kelp, and fish such as lumpfish thrive in pools on the lower shore.

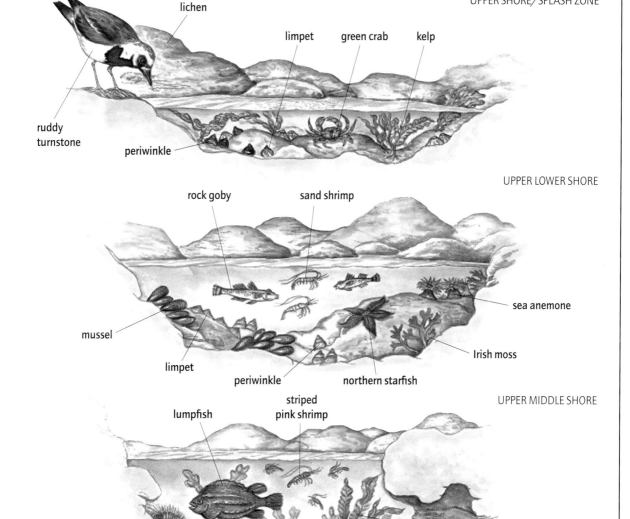

UPPER SHORE/SPLASH ZONE

UPPER LOWER SHORE

UPPER MIDDLE SHORE

WATERY WORLD

TRY THIS

Build a Minipond

Build a minipond in your backyard using an old washbowl or a piece of plastic sheeting. First, ask your parents to help choose a good site for the pond. Dig a bowl-shaped hollow with a spade or trowel, place the bowl or plastic sheeting inside, and then firm the soil back around the edge. Cover the bottom with a layer of gravel, and then put stones around the edge and on the bottom. Add water plants, and fill the pond with water. It won't be long until wildlife begins to colonize (move into) the pond.

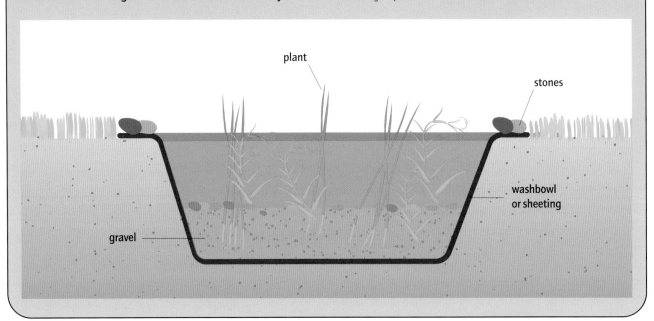

Space the water plants evenly through your garden minipond to encourage aquatic wildlife to colonize the new habitat.

Freshwater ecosystems

Freshwater covers only a small part of Earth's surface but offers a range of habitats. Freshwater ecosystems may be divided into still-water systems, such as lakes, ponds, swamps, and marshes, and habitats with flowing water, namely, rivers and streams. As in the oceans, the productivity of freshwater habitats is affected by many factors. These factors include soil type, water depth, water temperature, sunlight and oxygen levels, minerals, and pollution. Freshwater habitats rich in nutrients and organic matter are also rich in life. Wetlands fed by water from acidic rocks or peaty ground are particularly low in nutrients.

Oxygen levels and pollution

The levels of oxygen in freshwater habitats are affected by pollution. Even mild levels of pollution in a habitat can have a harmful effect on local wildlife. Fertilizers used in farming to improve crop yield contain many different nutrients. Water drained from farmed fields makes freshwater systems superrich in nutrients, which then causes algae and bacteria to multiply. Algae blanket the water surface and, as a result, block sunlight from plants that produce oxygen. When the algae die, oxygen-consuming microorganisms eat the algae and use up lots of oxygen. As a result, fish and other wildlife suffocate and die, and the ecosystem is completely altered.

ECOLOGY

USES OF WETLANDS

Rivers, lakes, and other bodies of freshwater have countless uses. Most important, they supply water for wildlife and also for human settlements, farming, and industry. Around the world many settlements are located by rivers, lakes, and other wetlands or on the coast. Wetlands act as reservoirs and also help prevent flooding after heavy rain has fallen by soaking up excess water. They also provide a wealth of food from fish, shellfish, and waterfowl to fruit and crops such as rice. Aquatic mammals such as otters and muskrats once yielded valuable fur for clothing. Timber and materials like rushes, traditionally used in roofing, also come from wetlands.

SCIENCE WORDS

- **countershading** Organisms with a dark top surface and a lighter lower surface are said to be countershaded; common in creatures that occur in surface waters, like penguins and plankton.
- **euphotic zone** Upper layer of the ocean that receives sunlight.
- **hydrothermal vent** A crack in rocks deep under the sea from which streams of very hot, chemical-laden water well up from inside the Earth's crust.
- **salinity** The salt content of water.

Mangrove trees line a riverbank in Thailand, forming a unique habitat that supports a range of aquatic organisms.

HUMAN IMPACT

Over the last few centuries, people have spread to every part of Earth and transformed many landscapes. As human numbers increase, we put more and more pressure on the natural world.

Modern human civilization began about 9,000 years ago as people began to grow crops in parts of the Middle East and, somewhat later, Egypt, northern India, and China. Settlements grew up in places where people could find food and water, and also safety from their enemies. Farmers learned to irrigate the land so crops could be grown. New techniques and machinery gradually made farming ever more efficient. Now many modern crop fields are "green deserts" where just one plant, the crop, is nurtured at

LIFE IN THE CITY

Many cities founded just a few centuries ago are now huge. The world's biggest cities include Mexico City, the capital city of Mexico; Sao Paolo in Brazil; Seoul, the capital city of South Korea; Tokyo, the capital of Japan; Moscow, the capital of Russia; London, the capital of Britain; and New York in the United States. In overcrowded cities hygiene, sewage management, and even getting clean water may be a problem, particularly in developing countries. More than half the world's population now live in cities.

Sao Paulo in Brazil is one of the most populous cities in the world, with more than 11 million inhabitants.

ECOLOGY

RAIN FOREST DESTRUCTION

Tropical rain forests contain the greatest biodiversity of any habitat on Earth. Experts estimate that around two-thirds of all species live there. However, this precious habitat is also disappearing faster than any other. Tropical rain forests now occupy less than half the area they did just a century ago. These forests are being felled for their valuable hardwood timber or for fuel and to clear land to grow crops and ranch cattle. Mining and dam building are also nibbling away at forests in some areas. This problem is called deforestation.

Destruction of rain forests harms the balance of gases in the atmosphere and affects the world's climate. Forest trees absorb carbon dioxide and so increase oxygen levels in the atmosphere. Felled trees can no longer take in carbon dioxide or release oxygen.

In addition, if the forests are burned to clear the land, carbon is released, adding to levels of carbon dioxide in the atmosphere. Since carbon dioxide is a greenhouse gas, deforestation contributes to the problem of global warming—rising temperatures worldwide.

This photograph has been taken from an airplane and shows a vast swathe of prime Amazonian rain forest that has been cleared for timber.

the expense of wild plants and insect life, which farmers poison with **pesticides** and **herbicides**.

From the 1790s, the Industrial Revolution saw the start of modern manufacturing methods. Now factories are built all over the world. Modern industries use up huge amounts of energy and natural resources, and produce considerable pollution. And in the last 20 years the human population has mushroomed. In 1980, there were about 4 billion people on Earth; by 2000 there were 6 billion. Experts think there may be 9 billion by 2050. This puts a huge strain on natural resources and the environment.

Habitat destruction

One of the main effects of the growing number of people on Earth is habitat destruction, as people take over the natural habitats where wild species once lived. Every year more and more wild land is plowed just to grow enough food for people or to graze domestic livestock. Overgrazed land is prone to erosion; the animals eat or trample everything, including the roots that hold together the soil. Huge areas of land also disappear under concrete as new towns and industrial parks spring up, and roads, railroads, and powerlines are blazed through forests, deserts, and wildernesses.

HUMAN IMPACT

The Okavango Delta in Botswana is the world's largest inland delta. It forms where the Okavango River empties into a swamp in the Kalahari Desert. Climate change is now threatening this fragile environment.

In many parts of the world, governments are handing over wild land to mining companies to dig or drill for minerals and fuel, or to construction firms to build dams to produce hydroelectric power. The tourist industry also puts increasing pressure on wild areas such as mountain regions, coral reefs, and sandy beaches. Many people like to "get back to nature" on vacation, yet sprawling resorts can spoil the scenery that tourists come to enjoy. Large numbers of visitors can disturb wildlife such as turtles nesting on beaches.

VANISHING WETLANDS

Around the world natural wetlands are disappearing fast. They are being drained to provide land for farming or new sites for factories and towns. Rivers and other freshwater systems are mined for gold and gemstones, building materials such as sand and gravel, or fuels such as peat. Some wetlands also disappear as they dry or silt up naturally, or when the land subsides. The United States has lost half its wetlands since the start of European settlement. Two-thirds of Europe's wetlands have vanished in the last 300 years.

ECOLOGY

WHALING

The European whaling industry began around 1600. Throughout the 1700s, 1800s, and early 1900s whales were hunted relentlessly for their meat, oil, bones, and baleen (hornlike material on the upper jaw). By 1950 many of the once-numerous great whales were almost extinct. In the 1990s, some whaling was banned by many nations. Now that the slaughter has ended, many whale populations have begun to recover and grow again.

Overhunting

As human numbers have risen, overexploitation of the world's wildlife has become a major problem. For thousands of years people have hunted animals on land and in rivers and oceans for meat and also for hides, fur, or feathers. When the human population was smaller, this hunting did little harm. In recent centuries, the invention of guns, explosive harpoons, and devices such as sonar, used to detect fish, have made the killing much easier. Overhunting and overfishing now threaten the survival of many ocean and land species.

In addition to hunting for food, dangerous animals such as tigers, sharks, and snakes are often hounded because people fear them as killers. Elephants, rhinoceroses, and big cats are killed, often illegally, for their ivory or fur. Elsewhere the pet trade threatens animals as diverse as monkeys, parrots, and terrapins. Animals that are captured in the wild rarely survive long in captivity. In the past, overhunting killed off species such as the dodo. Ecologists fear that many more species may soon go the same way.

Introduced species

Every ecosystem on Earth has its own community of plants, animals, and other organisms suited to that particular environment. The **introduction** of new species to an area can upset the natural balance of life there. All over the world people

The cane toad is a large, poisonous toad native to Central and South America. It has been introduced to many countries, often with disastrous consequences.

UNWELCOME ADDITION

In the 1930s a panel of experts recommended that the cane toad of Central America should be introduced to Australia to control beetle pests in the sugar cane plantations there. These amphibians were duly brought to Australia and released. Unfortunately, the large, poisonous toads bred quickly and now threaten native frogs, reptiles, and even small mammals.

HUMAN IMPACT

TRY THIS

Monitoring Water Pollution

The presence of certain aquatic creatures in streams and rivers gives clues about the water quality. Caddisfly and mayfly larvae and freshwater shrimp thrive only in water that is clean. See if these creatures can be found in a stream or pond near you. Take care not to get too close to the edge and fall in. Use a net to sweep the water, and then transfer your catch to a bowl while you identify the species using a field guide. Return the minibeasts to the water after you have finished looking at them.

have brought in new, non-native plants and animals to provide food, to control native species seen as pests, or simply to improve the scenery. If the new arrivals then thrive and multiply in number, they can quickly crowd out and threaten the original native species living there.

Pollution

Pollution means any form of contamination of land, air, or water often caused by people. It may be deliberate or happen accidentally. Pollution is spread though the air by winds, by currents in moving water, and through the soil by seeping

Adult caddisfly such as the one below hatch from aquatic larvae. The presence of caddisfly larvae in a body of water is a good indicator that the water is clean.

ECOLOGY

groundwater. Once in the ecosystem the contamination is absorbed by wildlife at a point in the food chain and passes up through the chain as carnivores eat herbivores. Top predators that feed on large numbers of polluted prey are sometimes the most contaminated of all. In the early 1960s, for example, the American bald eagle nearly died out because pesticide concentrations built up in its prey and thus within the eagles' own bodies. As a result, their eggshells were damaged and became too fragile for the eggs to survive.

Human activities such as mining, manufacturing, energy production, and farming produce substantial amounts of pollution. So does the waste we produce daily in our homes, offices, and schools. Fossil fuels (oil, gas, and coal) burned in factories, power plants, and cars release toxic fumes into the atmosphere. Car exhausts spew out carbon dioxide, carbon monoxide, and sometimes lead. Devices called catalytic converters can reduce this pollution. These poisonous gases are a major cause of the dirty smog that hangs in the air over cities, causing breathing problems.

Chemicals from factories and fertilizers and pesticides used in farming pollute rivers, lakes, and oceans. The soil is poisoned by chemicals from industry, mining, and agriculture, and also by the disposal of waste in sites called landfills. The problem of how to deal with all the waste we produce in our homes and factories gets bigger each year.

Acid rain is a form of pollution that results when nitrogen oxide from car exhausts and sulfur dioxide from power plants enter the atmosphere. These gases mix with water vapor in the air to form weak acids, which then fall as rain. Acid rain can kill trees, harm wildlife, and eat away at buildings.

Ocean pollution

For centuries oceans have been used as a dumping ground for sewage and all types of dangerous chemicals. A staggering 20 million tons of rubbish

The bald eagle nearly became extinct following the introduction of a pesticide called DDT in the mid-20th century. The species has since recovered.

CLEANING UP THE COMMUNITY

Here are some of the measures agriculture and industry are taking to reduce pollution in our community:

- Farmers are using natural fertilizers, such as manure, instead of factory-made chemical fertilizers.
- Companies are making heating systems with self-regulating thermostats in order to save energy.
- Traditional incandescent light bulbs are being replaced with energy-saving fluorescent light bulbs.
- Stores are providing reusable plastic bags to carry shopping.
- Recycling centers can process a range of different materials, from plastics and printer cartridges to batteries and beverage cartons.
- Farmers are planting drought-resistant native plants that require less water and fertilizers than exotic crops.
- Companies are producing environmentally friendly cleaning products to replace harmful detergents.

HUMAN IMPACT

THE CHERNOBYL DISASTER

In 1986, one of the world's worst-ever pollution disasters was caused by an accident at a nuclear power plant. The nuclear reactor at Chernobyl in Ukraine, part of the former Soviet Union, caught fire and exploded. A huge cloud of radioactive gas escaped and spread over a vast area. In Scandinavia and Siberia tundra vegetation was contaminated, which poisoned wildlife such as reindeer (caribou). Thousands of deer had to be destroyed.

The number of polar bears is in decline thanks to global warming and the consequent loss of habitat, as well as pollution from mining activities in the Arctic.

are dumped in the oceans every year, including huge items of junk such as discarded oilrigs. Coastal waters are often the most heavily polluted because pipes discharge raw sewage and chemicals into the shallows. Filter feeders and scavengers such as mollusks and crustaceans absorb the pollution and are then eaten by fish and other predators. Researchers in the Arctic have monitored pollution levels in top predators such as polar bears that feed on fish and seals. Some of the bears contained so much poison they could be classed as toxic waste.

Pollution-related hazards

Pollution of the air, water, and land has led to a number of very serious environmental problems. One of them is global warming, caused by the increase in greenhouse gases in the atmosphere partly due to burning fossil fuels. The warmer weather causes ice in the polar regions to melt. If this continues, rising sea levels may threaten low-lying coastal regions, such as Louisiana, Bangladesh, and the Netherlands. Ozone loss in the atmosphere,

POLAR POLLUTION

The polar regions are generally far from major towns and industrial centers that release pollution. However, they are rich in minerals that are mined, which can cause contamination of both the land and the sea. In the Arctic mining and drilling for gold, uranium, lead, coal, oil, and natural gas have caused considerable pollution on land and in the water. The Antarctic is also rich in minerals, but the continent is a wildlife sanctuary, so no mining is allowed there. However, traces of pollution from distant sources, such as radioactive contamination from Chernobyl, have been found in the Antarctic ice.

ECOLOGY

GREENHOUSE GASES

The five natural greenhouse gases are water vapor, carbon dioxide, methane, nitrous oxide, and ozone. The atmosphere also contains human-made greenhouse gases called chlorofluorocarbons (CFCs). As the level of greenhouse gases increases, Earth's climate changes—an effect called global warming.

Earth's atmosphere traps more heat if there are more greenhouse gases. Some of the heat escapes, but most of it is trapped, warming the planet.

caused by polluting chemicals called CFCs, means that more harmful ultraviolet radiation reaches Earth.

Oil spills

Oil spills are a major source of marine pollution. One of the worst-ever oil spills affected a huge area of Arctic coast. In 1989, the Exxon oil tanker *Valdez* ran aground off Prince William Sound in Alaska, spilling 50,000 tons (45,000 t) of oil into the sea. More than 1,200 miles (1,900 km) of coastline were polluted, and thousands of birds, sea otters, fish, and shellfish died. The disaster cost billions of dollars to clean up.

Nuclear hazards

Since burning fossil fuels releases chemicals that pollute the air, scientists are looking for other ways to generate the electricity we need to light and heat our homes and power our automobiles and airplanes. Nuclear power is becoming popular in many countries. Nuclear power harnesses the energy locked up inside atoms to generate electricity.

One of the main advantages of nuclear power is that it doesn't produce harmful gases that pollute the atmosphere. And unlike fossil fuels such as coal and natural gas, there is also a limitless supply of uranium fuel to use in nuclear power plants.

However, nuclear power comes with its own problems. The biggest drawback is what happens to all the radioactive waste produced by the power plants. Nuclear waste produced today will remain radioactive for the next 500,000 years. Each year, nuclear power plants in the United States are producing thousands of tons of waste. It costs a lot of money to store nuclear waste, and there are not enough disposal sites. As a result, a lot of waste is lying in temporary storage at the power plants.

Another problem is what happens if there is an accident. One of the worst environmental disasters occurred when harmful radiation escaped from the nuclear reactor at the Chernobyl power plant in the

HUMAN IMPACT

TRY THIS

Saving Energy and Recycling

We can all tackle the problems of global warming and acid rain by saving energy and recycling household waste. Think about ways in which your family or school could save energy. Switching off lights and machines when they are not needed, turning down the central heating, and making compost from vegetable peelings can help. So can using the car less. Note all the car journeys your family makes in a week. Are they all necessary, or could you walk, cycle, or take public transportation instead? Taking old cans, bottles, paper, and plastic packaging to recycling centers so they can be used again reduces waste. Use fewer chemicals and fertilizers on lawns.

Ukraine (see the box on page 52). Since radioactive waste is so harmful and long lasting, and the effects of radiation leaks are so damaging, many environmental organizations are calling for a ban on nuclear power.

Minimizing human impact

What can be done to reduce pollution and other damage to the natural world? Creating sanctuaries and reserves where no development is allowed can reduce the effects of habitat destruction. Building on brown field sites—empty ground in city centers—can preserve wild habitats on the edge of towns.

Replanting can restore lost forests, and careful management of land can prevent erosion. Organic farming, in which farmers avoid chemical fertilizers and pesticides and rely on traditional methods such

We can all do our bit to help protect the planet. Why don't you help your parents around the house by recycling trash such as aluminum cans, plastics, and glass bottles?

ECOLOGY

ECOTOURISM

In many parts of the world **ecotourism** now helps protect wildlife. Ecotourists pay to see wildlife in natural surroundings such as reserves. Their money helps pay for the park upkeep and for local conservation programs. Ecotourism helps the conservation work in reserves worldwide, such as Australia's Great Barrier Reef, in the wilds of Antarctica, and in Nepal's Annapurna mountain sanctuary.

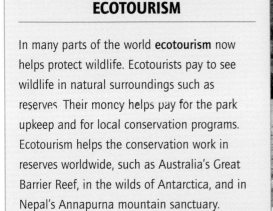

as crop rotation, can minimize the impact of farming on the environment. The major cause of pollution and habitat destruction is the human population explosion. It can be curbed by more effective birth control.

Governments around the world are introducing laws to control the release of chemicals by cities, industry, and agriculture. Cooperation is needed to tackle ozone loss and global warming. Many countries no longer use CFCs, which cause ozone loss. Governments have met to agree on targets to reduce greenhouse gas emissions. Development of renewable energy sources such as solar, wind, and water power can slow down global warming. People can tackle pollution by using less energy and chemicals and reducing waste.

Many people like to visit wilderness areas on vacation and experience the natural world at first hand. The money generated from this ecotourism can help pay for conservation programs.

SCIENCE WORDS

- **ecotourism** Vacations in which the natural wildlife and landscape are the main attractions for the holidaymakers.
- **herbicide** Chemical that kills off pest plants such as weeds.
- **introduction** The establishment of a species into a new area, sometimes through the activities of people.
- **pesticide** Chemical that kills pest organisms.

CONSERVATION

Conservation is the action taken by people to protect and preserve the natural world and its resources, such as living organisms, to prevent them from going extinct and losing valuable species forever.

Generally, people agree that natural resources should not be used up or wasted, but they may have different views about what makes a good conservation plan for a particular region.

> **THE AMERICAN BISON**
>
> Millions of American bison, or buffalo, once roamed the open plains of North America. Many Native American peoples depended on them for food, clothing, and shelter. In 1830, the government began killing the bison to control the Native Americans. By 1900, fewer than 1,000 bison were left, and conservationists worked hard to save them.

People often disagree about the right balance between environmental interests and economic ones. Sometimes scientists think that a habitat should be preserved and protected but differ on how people should use the area. For example, someone who visits a national park to relax or enjoy listening to animal sounds does not appreciate someone else who uses a snowmobile or motorized all-terrain vehicle there. People who participate in bird watching or hiking activities do not want to do so at the same time that others are hunting wild game. A major role of conservation biologists is to settle such conflicts.

Conservationists believe it is very important to look after the land and water as well as living organisms. Sometimes the best conservation plan to protect a habitat is to limit the number of people allowed in and let the ecosystem operate naturally.

Basic principles

Conservation biologists consider three basic principles for developing a conservation plan. First, that evolutionary change is a basic process in any living system. All species in natural ecosystems

Conservation work can be as varied as helping endangered animals in their natural habitat to teaching local people how to be more environmentally responsible.

ECOLOGY

HOW DID CONSERVATION BEGIN?

Several centuries after the colonization of North America by Europeans people began to realize that natural resources were limited and could be used up by activities such as clearing forests and building towns, cities, and factories. In the 19th century the writer Henry David Thoreau (1817–1862) promoted an idea called the romantic-transcendental conservation ethic (a harmonious appreciation of and coexistence with nature). During the same century the naturalist John Muir (1838–1914) helped form the Sierra Club, which encouraged people to enjoy, learn about, and preserve natural settings. His efforts contributed to the creation of several national parks, including Yosemite.

The resource conservation ethic, promoted by the forester Gifford Pinchot (1865–1946), was based directly on economics and commercial interests. Pinchot believed that natural resources, including forests, should be used to produce materials for human use. This viewpoint led to a type of forest management in which commercial markets were built into the conservation plan so that timber production, mining, hunting, and fishing were balanced against habitat and wildlife preservation.

The most scientific approach to conservation, the evolutionary-ecological land ethic, was developed by the naturalist Aldo Leopold (1887–1948) in the early 20th century. Leopold viewed natural ecosystems as complex, interworking parts. Most scientists in conservation programs support many of these views. Modern approaches to conservation combine the three ethics identified above.

The world's population of tigers is in serious decline and two of the six surviving subspecies are critically endangered. International laws have been agreed to protect those that are left in the wild.

respond to environmental change, so conservation plans must consider how species can respond based on their history. Long-term conservation plans that involve changing and managing habitats need to consider differences in the life styles of various living organisms as a consequence of their evolutionary past. Many insects, small mammals, and native

57

CONSERVATION

Yellowstone National Park is a protected wilderness area in Wyoming, Montana, and Idaho in the United States.

grasses live no longer than a year, while freshwater turtles, tortoises, and oak trees grow for several years before becoming adults, maturing and then producing offspring.

The second principle is that the environment is always changing because of factors such as weather, season, and species interactions. Modern conservation biologists do not expect the pattern of distribution and the number of plant and animal species in an area to remain the same. This principle goes against an earlier concept that nature is in a balanced state.

Conservation plans should be based on an understanding that the types of species and their population levels constantly evolve and change during the course of time. For example, if beavers move into an area with a small stream, they usually build dams. The dams then create ponds that might attract some species, such as wading birds. However, decreasing the stream flow in this way might result in there being a reduction in certain other species, such as streamside salamanders.

The third and final principle of any conservation plan is human society. Human needs and preferences can easily cause conflict and disagreement. There are situations in which conservation biologists must think about the needs or desires of the original inhabitants of a region. For example, conservation plans to preserve and protect natural habitats in Australia

THE GAIA HYPOTHESIS

In the 1960s, British scientist James Lovelock (born 1919) published the **Gaia hypothesis**. It is the idea that Earth's oceans, atmosphere, and landmasses are held in equilibrium (balance) by all the living organisms of the planet, including humans. The hypothesis is that the living world functions as a superorganism, keeping a worldwide environmental balance. However, not all scientists agree with the idea.

ECOLOGY

now include the attitudes and feelings of the aboriginal peoples, who are the original Australian inhabitants. Also, the International Whaling Commission (IWC) has a quota (allowance) for the hunting of bowhead whales. This permits Alaskan Inuits and the native people of Chukotka, Russia, to follow their cultural traditions of whale hunting.

Ecological reserves

One of the greatest threats to biodiversity and ecosystems is the destruction of habitats. Conservation programs have resulted in the establishment of ecological reserves that protect species from human activities and allow them to live in these regions. Many countries have put aside land, freshwater, seashore, and marine habitats as national parks and wildlife refuges for the protection of living organisms.

One of the most successful U.S. conservation approaches ever taken was the setting aside of natural lands as national parks. Yellowstone National Park in the western United States, established by Congress in 1872, was the world's first national park. Royal National Park near Sydney, Australia, established in 1879, became the second.

Many countries in Africa, Europe, Asia, and tropical America have also used the national park concept of conservation, so now hundreds of protected habitats exist around the world.

Marine conservation

Conservation programs for marine mammals are very important worldwide and often involve international politics. The International Whaling Commission (IWC) was established in 1946 to protect whales from overhunting and achieve the best population levels of commercial species. In this way whale hunting could carry on without all the whales being killed. However, Norway and Japan, countries that have always hunted whales, have exceeded IWC set quotas many times.

 TRY THIS

Backyard Habitat

A basic principle of conservation programs is that more natural habitats mean more native wildlife. Large ecological reserves are often part of regional conservation plans, but you can see the positive effects by creating a habitat on a much smaller scale, such as in your backyard. To develop and maintain a backyard habitat, you need four basic things—food, water, shelter, and places for animals to have young. You also need a mixture of shrubbery, trees, and open areas to make an ecologically healthy and diverse backyard habitat that reinforces the importance of creating ecological reserves.

Road kill

Loss of habitat is one of the greatest threats to animals in most countries, and highways are a major cause. In the United States highways take up around 1 percent of the land surface, and about one million animals are killed on the roads every day. By noting the number of road-killed animals in an area, particularly after a highway has been newly built, conservationists can find out how badly natural habitats are affected and help make people aware of the importance of finding ways to prevent as many animals as possible from being killed in this way.

Captive-breeding programs

When an animal species is threatened by extinction, conservationists may collect some of the animals and breed them in captivity. Many **captive-breeding** programs have been successful, including those with the peregrine falcon. In the 1950s and 1960s many farmers used pesticides to improve

59

CONSERVATION

their crop yields, but the chemicals built up in many types of wildlife in the food web. Because they fed on contaminated prey, many birds saw their eggshells become fragile. The survival of young peregrine falcons declined, greatly reducing their numbers. In 1960 no breeding pairs were known to exist in the eastern United States. Ten years later a captive-breeding program was started using peregrines from the western United States and Europe. The program was so successful that in 1974 conservationists released the first peregrine falcons back into the wild. Now peregrine falcons breed successfully in many of the places in which they once lived.

Although some endangered species in zoos and other facilities have been bred successfully, then

TURTLES AT RISK

Marine turtles are at risk of becoming extinct. There are several causes for this, including disruption of their breeding by human interference. Some people have collected and eaten the turtle eggs they find near beaches; others have made turtle soup from adult green turtles. In Australia red foxes have preyed on turtle nests. Even though there are laws to protect the marine turtles, their population levels are still very low. In some places turtle eggs are taken indoors until they hatch, then the young turtles are released. Do you think this will solve the problem of marine turtles becoming extinct?

The loggerhead turtle is an endangered species. Many turtles get caught up in fishing nets and drown. Habitat loss through human activities such as building marinas and dredging are also decimating the population.

ECOLOGY

released back into the wild, some serious problems with the programs have emerged. Any excessive hunting, pollution, or habitat destruction that put the species in danger in the first place still exists in most cases. It turns out that simply adding more animals to a troubled habitat is not an effective conservation measure.

Impacts of explorers

Without conservation plans to prevent the exploitation and destruction of natural resources even small numbers of humans can wipe out species simply by moving into their native habitats and outcompeting them.

Some conservation biologists think that explorers and settlers throughout the world have wiped out countless species that were easily preyed on by people. The Polynesian settlers called Maoris, for example, killed off a group of flightless bird called the moa in New Zealand. These birds were among the largest ever known, some standing more than 10 feet (3 m) tall. They were not adapted to avoid human hunters; they had no defenses against the settlers and could not fly away from danger. Most species of moas were extinct within 100 years of the settlers' arrival.

Panda protection

Scientists monitored the effectiveness of setting aside habitats to protect wildlife from human activities at an ecological reserve for giant pandas in China. They found that the panda habitat had declined in quality and decreased in size since the establishment of the reserve in 1975. A rapid increase in the human population within the reserve, which led to increased tourism, removal of wood, and road construction, was thought the main cause of the habitat loss. The study reinforced the principle that effective wildlife conservation plans must always include human factors, activities, and needs.

A giant panda feeds on bamboo shoots in a forest in China. Loss of habitat is one of the main reasons for the decline in the panda population.

SCIENCE WORDS

- **captive breeding** The process of taking wild animals and mating them in captivity to increase the species under protected conditions.
- **conservationists** People who try to preserve the natural world.
- **Gaia hypothesis** Idea that Earth functions as an enormous "superorganism" that maintains the conditions necessary for its survival.

GLOSSARY

biodiversity The total diversity of organisms and genes in a given area.

biomass The total weight of all the organisms in an area (or trophic level).

biome A type of major ecological community, such as a tropical rain forest, desert, or tundra.

biosphere All the parts of Earth that are able to sustain life; includes the lower atmosphere, oceans, and the surface of the land and fresh waters, and extends to a mile or so below the surface.

brackish Semisalty water, as in estuaries and deltas.

canopy Uppermost level at the tops of trees in a forest.

captive breeding The process of taking wild animals and mating them in captivity to increase the species under protected conditions.

carbon cycle Cycle of carbon through the natural world.

carnivore An animal that eats meat.

cellulose Tough chemical that forms part of the cell walls of plants.

CFCs Chemicals called chlorofluorocarbons used in the manufacture of refrigerators, polystyrene, and in aerosol spray cans; a major depleter of the ozone layer.

cladistics Technique that compares large numbers of characteristics among species to build up a family tree.

classification The organization of different organisms into related groups by biologists.

climate The regular weather pattern that occurs in a certain region.

community A group of different species that share a habitat.

conservationists People who try to preserve the natural world.

countershading Organisms with a light lower surface and a darker top surface are said to be countershaded; common in creatures that occur in surface waters, like penguins and plankton.

dark zone Area of the deep sea to which light from the surface cannot reach.

delta Land crossed by small water channels that forms around the mouth of a river through the laying down of sediment.

ecosystem An ecological unit that comprises a community of organisms and its environment.

ecotourism Vacations in which the natural wildlife and landscape are the main attractions for the holidaymakers.

euphotic zone Upper layer of the ocean that receives sunlight.

food chain The passage of energy between organisms; a plant links to a herbivore, which in turn links to a carnivore. Energy is lost with each step.

food web A complex series of interlinked food chains.

fossil fuel Carbon-based fuel, such as oil or coal, that forms from the remains of ancient organisms.

Gaia hypothesis Idea that Earth functions as an enormous "superorganism" that maintains the conditions necessary for its survival.

greenhouse effect The rise in global temperatures caused by an increase in gases such as methane and carbon dioxide in the atmosphere.

habitat The type of place in which an organism lives.

herbicide Chemical that kills off pest plants such as weeds.

herbivore Animal that feeds on plants.

hydrothermal vent A crack in rocks deep under the sea from which streams of very hot, chemical-laden water well up from inside the Earth's crust.

introduction The establishment of a species into a new area, sometimes through the activities of people.

mesosphere Atmospheric level between troposphere and stratosphere containing little water vapor, and where the temperature drops to –112°F (–80°C).

microhabitat A small part of a habitat that sustains a community; for example, a pool in the leaves of bromeliad plants forms a microhabitat.

natural selection Theory that only the organisms best suited to their environment survive to reproduce. Natural selection is the driving force behind evolution.

niche The ecological role of an organism in a community.

nitrogen fixation The incorporation by soil bacteria of nitrogen in the air into nitrate compounds that plants are able to use.

organelle A membrane-lined structure inside eukaryote cells, such as the nucleus.

ozone layer Layer of ozone gas high in the atmosphere. It filters out harmful ultraviolet radiation from the Sun.

pesticide Chemical that kills pest organisms.

photosynthesis The conversion of water and carbon dioxide into sugars in plants, using the energy of sunlight.

phytoplankton Plantlike algae that float in the surface waters of lakes or the ocean.

predator Animal that catches other animals for food.

prey Animal caught and eaten by another animal.

protist A single-celled eukaryote organism with a nucleus and organelles.

respiration The process of reacting food with oxygen to liberate energy; takes place inside the mitochondria of eukaryotes or on the cell membrane of prokaryotes like bacteria.

salinity The salt content of water.

savanna A tropical grassland.

species A group of organisms that can potentially mate with each other to produce young that can also interbreed successfully.

stratosphere Upper level of the atmosphere. It contains the ozone layer; extends from the mesosphere to the edge of interplanetary space.

subspecies Subdivision of a species; a population that may have different colorings and a different range from other subspecies but can still interbreed with them.

territory Area controlled by an organism to protect food supplies or to attract a mate.

trophic level One of the links in a food chain; plants, herbivores, and carnivores are all at different trophic levels.

troposphere Lowest layer of the atmosphere. It contains most of the water vapor; clouds occur in this layer, and weather changes happen there.

tundra Cold, treeless near-polar region with a layer of permanently frozen soil just beneath the surface.

twilight zone Ocean level that receives only small amounts of light from the surface.

zooplankton Small animals that float in the surface waters of lakes or the ocean.

FURTHER RESOURCES

PUBLICATIONS

Bassett, J. *Science Activities: Our Environment.* Danbury, CO: Grolier Educational, 2002.

Cefrey, H. *What If the Hole in the Ozone Layer Grows Larger?* New York: Children's Press, 2002.

Giles, B. *Parasites and Partners: Lodgers and Cleaners.* Chicago, IL: Raintree, 2003.

Jennings, T. *Ecology.* Costa Mesa, CA: Saddleback Educational Publishing, 2009.

Latham, D. *Ecology.* Chicago, IL: Heinemann-Raintree, 2009.

Luhr, J. F. (ed). *Earth.* New York: DK Publishing, 2003.

Martin, J. W. R. *Small Worlds: In a House.* New York: Crabtree, 2002.

Morgan, B. (ed). *Biomes Atlases.* Chicago, IL: Raintree, 2010.

Morgan, S. *Earth Watch: Wildlife in Danger.* New York: Franklin Watts, Inc., 2000.

Nicholson, S. *Rainforest Explorer.* Lake Mary, FL: Tangerine Press, 2001.

Pringle, L. *Global Warming: The Threat of Earth's Changing Climate.* New York: SeaStar Books, 2001.

Riley, P. *Straightforward Science: Food Chains.* New York: Franklin Watts Inc., 2003.

WEB SITES

Edens
www.pbs.org/edens
Learn about the people, flora and fauna, and threats to some natural wonders.

Global Warming For Kids
www.epa.gov/globalwarming/kids/index.html
Facts, games, and climate animations.

Jaguar: Lord of the Mayan Jungle
www.oneworldjourneys.com/jaguar/index_flash.html
Join an animated expedition in the Yucatán jungle, and learn about jungle ecology and ecological preservation.

Madagascar: Biodiversity and Conservation
ridgwaydb.mobot.org/mobot/madagascar/default.asp
Learn about why Madagascar is such a special place and ecology, evolutionary biology, and threats to biodiversity.

Monterey Bay Aquarium
www.montereybayaquarium.com
Webcams, videos, field guides, activities, games, and habitat exhibits.

NASA Earth Observatory
earthobservatory.nasa.gov/Topics/life.html
Informative articles and spectacular satellite images show how NASA scientists are unraveling the mysteries of environmental change and many other ecological topics.

Voyage to the Deep
www.ocean.udel.edu/deepsea
Join a deep-sea expedition; with videos, photos, and multimedia tools.

Worldwide Fund for Nature
www.panda.org
Get the latest environmental news and read special features on protecting species and habitats.

Our editors have reviewed the Web sites that appear here to ensure that they are suitable for children and students. However, many Web sites frequently change or are removed, and we cannot guarantee that a site's future contents will continue to meet our high standards of quality. Be advised that children should be closely supervised whenever they access the Internet.

INDEX

Page numbers in *italics* refer to illustrations.

A
Algae 20, 39, 45
Alligators 6
Amphibians 12, 13, 49
 Cane toads 49
Antennae 9
Atmosphere, Earth's 18, 19, 20, 22, 25, 26, 27, 28, 29, 30, 31, 37, 50, 52, 53, 58

B
Bacteria 6, 8, 11, 12, 18, 19, 20, 21, 29, 41, 44
Bald eagles 16, 51
Biodiversity 7, 34, 39, 47
 Crisis 13, 59
Biomass 16, 17, 21, 39
Biomes 31
Biosphere 30, 37
Birds 4, 5, 7, 13, 14, 21, 24, 32, 35, 36, 37, 40, 53, 58, 60, 61
 Bald eagles 16, 51
 Gannets 36, 40
 Parrots 21, 49
  Macaws 21
 Vultures 15
Bison 56
Black panthers 10
Bones 14, 20, 49

C
Caddisfly 50
Cane toads 49
Captive breeding 59-60, 61
Carbohydrates 18
Carbon cycle 18, 21
Carbon dioxide 4, 7, 17, 18, 19, 25, 26, 27, 31, 47, 50, 53
Carnivores 7, 9, 11, 14, 15, 16, 17, 21, 36, 51
Caterpillars 4
Cats 10, 11, 14, 33, 37, 49, 57
 Black panthers 10
 Lions 11, 14, 33, 37
 Tigers 49, 57
Cells 8, 15, 18, 21
Cellulose 18, 21
Chernobyl disaster 52, 53-54
Cities 35, 46, 51, 55, 57
 Sao Paulo, Brazil 46
Cladistics 10-11, 15
 Cladograms 10, 11

Classification 8, 9, 10, 15
Climate 6, 7, 22-29, 31, 32, 33, 34, 36, 47
 Changing 27-28, 48, 53
 Zones 22
Conifers 25, 33
Conservation 56-61
 Workers 56
Coral reefs 8, 38, 41, 48
 Great Barrier Reef 41, 55
Corals 38, 40, 41

D
Darwin, Charles 14, 34
Deltas 38, 39, 48
 Okavango Delta, Botswana 48
Deoxyribonucleic acid (DNA) 15, 20
Deserts 5, 7, 22, 27, 31, 32, 34-35, 46, 47, 48
Digestion 16
DNA see Deoxyribonucleic acid
Dust bowl 32

E
Ecosystems 6, 7, 16, 17, 20, 30-37, 38-39, 41, 42, 44, 49, 51, 56, 57, 59
Ecotourism 55
Evolution 5, 10, 11, 14, 34, 56-57
Exoskeletons 9

F
Fertilizers 20, 44, 50, 51, 54
Fish 5, 11, 12, 13, 21, 33, 36, 38, 39, 40-41, 42, 43, 44, 45, 49, 52, 53, 57, 60
Fly agaric 12
Food chains 4, 7, 16, 17, 36, 39, 51
Food webs 4, 5, 7, 39, 42, 60
Forests 4, 5, 7, 31, 32-34, 39, 47, 54, 57, 61
 of Conifers 25, 33
Fossil fuels 18, 27, 51, 52, 53
Foxes, red 11, 17, 60
Fungi 6, 8, 12, 29, 31
 Fly agaric 12

G
Gaia hypothesis 58, 61
Gannets 36, 40
Global warming 18, 26, 27, 52, 53, 54, 55

Grasslands 7, 31, 32, 33, 37
 Savanna 22, 32, 33, 37
Greenhouse effect 26-27, 29, 53
Greenhouse gases 18, 47, 52, 53, 55

H
Habitats 5, 6, 7, 11, 35, 37, 39, 40, 41, 44, 45, 46, 47, 54-55, 56, 57, 58, 59, 60, 61
Hennig, Willi 10
Herbivores 7, 17, 21, 36, 37, 40, 42, 51
Hydrothermal vents 41, 45

I
Ice ages 27
Insects 4, 5, 8, 9, 10, 12, 29, 32, 35, 47, 57
Invertebrates 9, 12
Islands 34, 35

J
Jellyfish 39, 40

K
Köppen, Wladamir 22

L
Leaves 4, 7, 16, 18, 19, 33-34
Leopold, Aldo 57
Linnaeus, Carolus 8
Lions 11, 14, 33, 37
Lizards 5, 13
Lovelock, James 58

M
Macaws 21
Mammals 9, 11, 12, 13, 14, 24, 32, 36, 38, 45, 49, 57, 59
Mangrove trees 45
Mayr, Ernst 11
Meadows 16, 30
Microhabitats 7
Mountains 12, 24, 25, 31, 32, 48, 55
 Climate zones 25
 Kilimanjaro 24
Muir, John 57

N
Natural selection 5, 14
Nitrogen 19, 20, 25, 51
 Cycle 19, 20
 Fixation 19, 21
Nuclear waste 52, 53-54

O
Oceans 7, 8, 12, 21, 22, 25, 28, 30, 31, 37, 38-45, 49, 51-52, 58
 Zones 40
Oxygen 6, 7, 17, 18, 21, 25, 26, 28, 31, 38, 39, 44, 47
Ozone layer 25, 26, 29, 52, 53, 55

P
Palm trees 8, 22
Pandas 61
Parrots 21, 49
 Macaws 21
Pesticides 47, 51, 54, 55, 59
Photosynthesis 4, 16, 18, 39
Pinchot, Gifford 57
Plankton 18, 26, 31, 38, 39, 40, 45
Plants 4, 5, 6, 7, 8, 9, 11, 12, 13, 14, 16, 17, 18, 19, 20, 21, 22, 24, 25, 26, 28, 29, 30, 31, 32, 33, 34, 35, 36, 37, 38, 39, 41, 42, 43, 44, 45, 46, 47, 49, 51, 55, 58
Polar bears 5, 52
Polar regions 23, 28, 31, 32, 52
Pollution 20, 27, 41, 44, 47, 50-55, 61
Ponds 5, 6, 7, 20, 44, 50, 58
Populations 5, 11, 12, 13, 15, 35, 36, 37, 46, 47, 49, 55, 57, 58, 59, 60, 61
 Charts 36
Protists 8, 15

R
Rabbits 16, 17, 32
Rain forests 7, 22, 31, 33, 34, 39, 47
 Deforestation 47
Recycling 18, 20, 51, 54
Red foxes 11, 17, 60
Reptiles 11, 12, 13, 24, 32, 49
 Alligators 6
 Lizards 5, 13
 Turtles 40, 48, 58, 60
  Loggerhead 60
Respiration 18, 19, 21
Rock pools 42

S
Savanna 22, 32, 33, 37
Scrublands 31, 34

Sea anemones 38, 41, 42, 43
Sea level 23, 24, 26, 27, 52
Sea lions 15
Seashore zones 43
Seasons 23-24, 32, 34, 58
Sea urchins 42, 43
Seaweed 39, 40, 42, 43
Soil 6, 12, 19, 20, 21, 28-29
 Horizons 29
 Profiles 29
Species 5, 8, 9, 10, 11-13, 14, 15, 16, 34, 35, 37, 38, 39, 40, 41, 47, 49, 50
 Introductions 49-50
Spiders 4, 9, 12, 29
Sunflowers 16

T
Taiga 31, 33
Territories 36-37
Thoreau, Henry 57
Tigers 49, 57
Trophic levels 17, 21
Tundra 22, 25, 31, 32, 33, 52
Turtles 40, 48, 58, 60
 Loggerhead 60

V
Vultures 15

W
Walruses 15
Water 4, 6, 7, 17, 18, 19, 20, 22, 25, 26, 27, 28, 30, 31, 32, 33, 35, 37, 38, 39, 40, 41, 42, 44, 45, 46, 50, 51, 52, 56, 58, 59
 Collecting 27
 Water cycle 22, 28
Wetlands 38, 44, 45, 49
Whaling 49, 59

Y
Yellowstone National Park 58, 59